Systems Thinking, Critical Realism and Philosophy

Systems Thinking, Critical Realism and Philosophy: A Confluence of Ideas seeks to re-address the whole question of philosophy and systems thinking for the twenty-first century and to provide a new work that may be of value to both systems and philosophy. This is a highly opportune time when different fields – critical realism, philosophy of science and systems thinking – are all developing around the same set of concepts and yet not realizing it.

This book will be of interest to the academic systems community worldwide; due to its interdisciplinary coverage, it will also be of relevance to a wide range of scholars in other disciplines, particularly philosophy but also operational research, information systems and sociology.

John Mingers is Professor of Operational Research and Systems at Kent Business School, University of Kent.

Ontological explorations

Other titles in this series:

From One 'Empire' to the Next
Radha D'Souza

Science for Humanism
The recovery of human agency
Charles R. Varela

Philosophical Problems of Sustainability
Taking sustainability forward with a critical realist approach
Jenneth Parker

Dialectic and Difference
Dialectical critical realism and the grounds of justice
Alan Norrie

Interdisciplinarity and Climate Change
Transforming knowledge and practice for our global future
Edited by Roy Bhaskar, Cheryl Frank, Karl Georg Høyer, Petter Naess and Jenneth Parker

Conversations about Reflexivity
Edited by Margaret S. Archer

Relational Sociology
A new paradigm for the social sciences
Pierpaolo Donati

Sociological Realism
Edited by Andrea M. Maccarini, Emmanuele Morandi and Riccardo Prandini

The Economics of Science: A Critical Realist Overview
Volume 1: Illustrations and philosophical preliminaries
David Tyfield

The Economics of Science: A Critical Realist Overview
Volume 2: Towards a synthesis of political economy and science and technology studies
David Tyfield

Ontology Revisited
Metaphysics in social and political philosophy
Ruth Groff

Childhoods, Real and Imagined
Volume 1: An introduction to critical realism and childhood studies
Priscilla Alderson

Naturalizing Critical Realist Social ontology
Tuukka Kaidesoja

What's Critical About Critical Realism?
Essays in reconstructive social theory
Frederic Vandenberghe

Integrating Knowledge Through Interdisciplinary Research
Problems of theory and practice
Dominic Holland

The Contradictions of Love
Towards a feminist-realist ontology of sociosexuality
Lena Gunnarson

Systems Thinking, Critical Realism and Philosophy
A confluence of ideas
John Mingers

Systems Thinking, Critical Realism and Philosophy

A confluence of ideas

John Mingers

Routledge
Taylor & Francis Group
LONDON AND NEW YORK

First published 2014
by Routledge
2 Park Square, Milton Park, Abingdon, Oxon OX14 4RN

and by Routledge
711 Third Avenue, New York, NY 10017

Routledge is an imprint of the Taylor & Francis Group, an informa business

British Library Cataloguing in Publication Data
A catalogue record for this book is available from the British Library

Library of Congress Cataloging-in-Publication Data
A catalog record has been requested for this book

ISBN: 978-0-415-51953-3 (hbk)
ISBN: 978-1-315-77450-3 (ebk)

Typeset in Times New Roman
by Wearset Ltd, Boldon, Tyne and Wear

Printed and bound in the United States of America by Publishers Graphics, LLC on sustainably sourced paper.

Dedicated to my family – Julie, Laura and Emma

Contents

Figures

Tables

Acknowledgements

Portions of this book have appeared in other works of mine prior to its publication. I would like to thank the publishers for their permission to reproduce this material.

Chapter 2: 2004. 'Re-establishing the Real: Critical Realism and Information Systems Research', in J. Mingers and L. Willcocks (eds) *Social Theory and Philosophy for Information Systems*, Chichester: Wiley, pp. 372–406.
Chapter 3: 2011. 'The Contribution of Systemic Thought to Critical Realism', *Journal of Critical Realism* (10), pp. 303–330.
Chapter 6: 2002. 'Can Social Systems Be Autopoietic? Assessing Luhmann's Social Theory', *The Sociological Review* (50:2), pp. 278–299.
Chapter 6: 2004. 'Can Social Systems Be Autopoietic? Bhaskar's and Giddens' Social Theories', *Journal for the Theory of Social Behaviour* (34:4), pp. 403–426.
Chapter 7: 2006. *Realising Systems Thinking: Knowledge and Action in Management Science*, New York: Springer.
Chapter 8: 2008. 'Management Knowledge and Knowledge Management: Realism and Forms of Truth', *Knowledge Management Research and Practice* (6), pp. 62–76.
Chapter 9: 2006. 'A Critique of Statistical Modelling in Management Science from a Critical Realist Perspective: Its Role within Multimethodology', *Journal of the Operational Research Society* (57), pp. 202–219.
Chapter 10: 2009. 'Discourse Ethics and Critical Realist Ethics: An Evaluation in the Context of Business', *Journal of Critical Realism* (8:2), pp. 172–200.

I would like to thank Mark Johnson and Soren Brier for reading sections of the book and making constructive comments.

Part I

Foundations

1 Introduction

1.1 Introduction

In 1972, Ervin Laszlo published a seminal book called *Introduction to Systems Philosophy: Towards a New Paradigm of Contemporary Thought* (Laszlo 1972). This was a book that made a major contribution to systems thinking in the sense that it covered all the main philosophical questions – ontology, epistemology, cognition, ethics, metaphysics and so on. But it was also a book that aimed to make a contribution to philosophy. Coming out of the intellectual ferment of the 1960s, it argued that philosophy also needed systems thinking: that systems provided a new and vibrant synthetic approach to traditional philosophy, especially in contrast to the sterile impoverishment of the Western analytic tradition.

Unfortunately, for many years the call fell on deaf ears – mainstream philosophy, particularly in the UK and the US, carried on in its analytic and positivist form. But outside philosophy, especially within the social sciences, huge debates developed about the nature of science, particularly social science, in terms of a conflict between positivism and interpretivism/constructivism; the development of critically inspired, neo-Marxist positions; and finally post-structuralism and postmodernism, which threatened an end to philosophy almost as Fukuyama (1992) threatened an end to history.

However, writing this some 40 years later, in 2013, I believe that there are real signs that Laszlo's hope for a systemic philosophy may at last be being realized through developments within, and on the borders of, philosophy itself. The first of these developments, not yet perhaps within the core of philosophy but becoming extremely influential and important within the social sciences, is a philosophical approach known as critical realism. This began to develop in the late 1970s as a reaction against the irrealism and relativism of radical constructivism and, even more so, postmodernism. It is primarily associated with the name of Roy Bhaskar (1978; 1979; 1986; 1989) but many others have made contributions (Archer 1988; Archer 1995; Archer *et al.* 1998; Keat and Urry 1981; Sayer 1992). Interestingly, however, until quite recently Bhaskar has been only on the margins of mainstream philosophy probably because he has never been part of its institutions, either in terms of an academic position or journal publications. That is now changing with the increasing proliferation of his influence.

Put briefly, critical realism (as first developed) re-asserts the primacy of ontology over epistemology – that is, it asserts the existence of an independent, external world about which we may acquire knowledge, while recognizing the inevitably fallible and contextual nature of that knowledge. It draws on aspects of both positivism and interpretivism (as well as critical theory) but maintains a strongly realist and rationalist orientation in opposition particularly to post-modernism. What is of special interest for us, however, is that critical realism is deeply and fundamentally systemic in character. Although the main texts of critical realism make little reference to the traditional systems literature, the discourse itself, especially in its later 'dialectical' manifestation (Bhaskar 1993; Bhaskar 1994), is couched almost exclusively in systems terms.

The second development is within mainstream philosophy, especially within the philosophy of science. For many decades, under the aegis of positivism, the concept of explanation was tied to the idea of universal laws. Following Hume, observation would be made of constant conjunctions of empirical events; from these, universal laws would be hypothesized. Events were then 'explained' as deducible instances of the universal laws (the deductive-nomological model). It was not permitted to go beneath the surface of these event regularities and postulate underlying, and possibly unobservable, 'causes'. However, in more recent years the limitations of this approach have become ever more obvious and an alternative has been generating much interest and support. This approach eschews universal laws in favour of particular 'mechanisms' that causally generate the phenomena of interest to the scientist (Gerring 2007; Glennan 1996; Glennan 2002; Machamer 2004; Machamer *et al.* 2000; Salmon 1998a; Symons 2008). Apart from avoiding many of the problems besetting the deductive-nomological model, especially concerning induction, the idea of mechanisms fits much better with the actual practices of scientists (Bechtel and Abrahamsen 2005) and, as we shall see, with explanations in everyday life.

This explanatory mechanisms view of causation is also highly systemic in nature – mechanism is but another name for system – but, as with critical realism, the literature uses the terms but does not reference the systems thinking discourse explicitly (see especially, for example, papers by Wimsatt 1994; 2007). That said, these philosophical developments have been recognized by some systems theorists, e.g. Pickel (2004; 2007) and Bunge (2004a; 2004b). But there is another connection here, and that is that the concept of causal mechanisms is also central to critical realism although once again the connection does not seem to have been realized or recognized by members of either camp. Thus, we now have two developments in philosophy – critical realism and explanatory mechanisms – which both draw strongly on the systems tradition, and both share a common conceptual framework and thereby have the potential to begin to realize Laszlo's vision of a systems philosophy.

1.2 Developments in systems thinking

As well as these advances in philosophy, there have also been many developments within systems thinking itself (Mingers and White 2010). I would like to highlight three – the rise of second-order cybernetics and soft systems; the impact of Forrester's system dynamics; and the emergence of chaos and complexity theory.

1.2.1 Second-order cybernetics and soft systems

Cybernetics (meaning the study of self-governing mechanisms) is a particular branch of systems theory orginating in the 1940s (Heims 1993; Pickering 2002) and developed by scientists such as Norbert Weiner (1948), Ross Ashby (1956), Gregory Bateson (1973b) and Gordon Pask (1976). They were interested particularly in the way that systems could control themselves autonomously through the transmission of information within error-controlled feedback loops. This enabled them to explain the particular nature of living systems and also to explore how the brain and our cognitive processes worked. In studying, for example, the nature of perception[1] it became clear that what we perceive is no passive reflection of the external world but rather a very active construction of the human nervous system. Thus we have to recognize that, in principle, the observer is always part of the system being observed.

This insight developed into what became known as second-order cybernetics.[2] First-order cybernetics studies the mechanisms of the external world, while second-order cybernetics studies the process of observation itself. It is the *Cybernetics of Cybernetics* (Von Foerster 1975) or the study of *Observing Systems* (Von Foerster 1984) (where 'Observing' is to be read as both a noun and a verb), as two of the major works by Heinz von Foerster put it. Much of this work was carried out at the Biological Computer Laboratory (BCL) at the University of Illinois and reached its most developed form in the work of Maturana (1970a) and Maturana and Varela (1975; 1980) on 'autopoietic' systems – systems that produce or construct themselves (Mingers 1995b).

Autopoiesis is a fundamental concept that has radical consequences. In terms of physical molecular systems it defines the nature of life itself. Living systems are bounded networks of processes of biochemical production that produce the very components of which they are constituted. Although they interact with their environment, they are organizationally closed in the sense that they both produce, and are produced by, themselves. The concept of autopoiesis has been influential in a range of disciplines including law (Teubner 1987), sociology (Luhmann 1982b; Mingers 2002), management (Von Krogh and Roos 1994), computing (Winograd and Flores 1987) and literature (McGann 1991). Stemming from, but separate to, autopoiesis, Maturana and particularly Varela developed a theory of mind that was much more phenomenological than computational, conceptualizing cognition as non-representational and embodied (Maturana and Varela 1987; Varela *et al.* 1991).

At the same time, analogous developments were occurring in another area of systems – applied systems thinking or systems engineering. The systems approach was successfully being used in the design of complex engineering projects such as oil refineries, and methodologies for tackling these problems had emerged (Hall 1962). However, when these methodologies were applied to problems in human organizations they were not found to work well. The issue is that human beings are significantly different from machines and buildings. People, through self-consciousness and language, have the ability to conceptualize themselves and the systems that they are part of – they exist in a world of meaning and signification. This means that we cannot just take for granted, from the outside, the nature of a particular social system or social interaction but have to engage with the participants and become active observers.

This led to the development of an alternative systemic approach to problem-solving in organizations – what became known as 'soft systems thinking' as opposed to the 'hard systems thinking' of traditional engineering. This represents a similar paradigm shift to second-order cybernetics – problematizing the role of the observer/participant in systems analysis. It was most fully articulated by Checkland (1981; Checkland and Poulter 2006; Checkland and Scholes 1990) in a practical intervention approach called soft systems methodology (SSM) which, he argued, was underpinned by a phenomenological social theory (Husserl 1973, orig. 1913).

Both second-order cybernetics and SSM were mirroring changes that were occurring in social science more generally, as mentioned above. In particular, there was movement away from a positivist view of social science as little different from natural science towards one based on interpretivism and constructivism that argued, to varying degrees, that the social world was intrinsically very different from the material world and required a wholly different, hermeneutic approach. This led to a major dislocation within social science (and systems thinking) between what were seen as fundamentally incommensurable paradigms.[3] This in turn paved the way for critical realism to try to overcome or synthesize the fracture.

1.2.2 System dynamics

In general terms, system dynamics simply means the changing behaviour of systems, but in practice it has become associated specifically with the work of Jay Forrester at MIT, who has developed an approach to simulating the behaviour of large complex systems. Forrester was initially interested in the dynamic behaviour of whole industries such as supply chains (Forrester 1961) and of populations of people as in the growth and decay of cities (Forrester 1969). He identified the major flows of people, materials and money and the ways in which these were controlled through feedback loops. He then modelled these using systems of differential equations which were run on a computer to display the dynamic behaviour of the system over time.

The most ambitious model developed was called the 'world model' (Forrester 1973) and was actually a model of the whole world economy run to cover a period of 50 years (up to 2020). It involved the interaction of five major factors – population growth, food production, industrialization, natural resource depletion and pollution – and the results suggested that the rates of growth then being experienced were not sustainable because of lack of natural resources and the growth of pollution. The results were published in a Club of Rome report entitled *The Limits to Growth* (Meadows *et al.* 1972). This model was extensively critiqued, especially by economists who questioned the assumptions and disapproved of the modelling method, but in many ways the broad conclusions concerning natural resources and pollution have in fact been borne out (Turner 2008).

One of the problems with system dynamics originally was the lack of computer power and the poor graphics facilities, but with modern technology highly sophisticated graphical software[4] has been produced which makes both model development and the display of the results much more effective. System dynamics is now also used on a more micro level for exploring the 'mental models' that individuals have about how parts of their world work. This approach was popularized by Peter Senge in his book about the 'learning organization' called *The Fifth Discipline* (Senge 1990), where the fifth discipline is in fact systems thinking, and can be seen as a 'soft' version of traditional system dynamics.

1.2.3 Complexity theory

Complexity theory (Gell-Mann 1995; Kaufmann 1995) developed during the 1970 and 1980s in a range of disciplines – biology (Bawden 2007), chemistry (Prigogine and Stengers 1984), mathematics (Gleick 1988) and economics (Anderson *et al.* 1988). Traditionally, these 'hard' sciences have made a range of assumptions about the behaviour of systems in their domains that were increasingly found not to hold true. This led to a crisis which eventually resulted in the emergence of what was first called chaos theory and later complexity theory. Before this, the assumptions had been mainly that the types of behaviour displayed were generally orderly and fairly predictable: for example, that systems were usually stable and reached equilibrium; that changes tended to be linear or at least smoothly non-linear; that systems exhibited cyclicality and robustness; that simple models would generate simple behaviour (and vice versa).

Chaos and complexity theory showed, both through models and in actual experiments, that systems could display instability, far-from-equilibrium behaviour, sudden and often inexplicable change, and extreme sensitivity to initial conditions, and that simple models could display complex and chaotic behaviour (and vice versa) (Lewin 1992; Mainzer 1997). Two important questions are: to what extent do these insights apply to the soft sciences and organizations (Byrne 1998; Cilliers 2000)? And, to what extent can complexity theory be encompassed within traditional systems thinking?

Complexity theory has been taken up enthusiastically within management and organization theory. We can distinguish three broad areas:

- An organization's environment is complex, characterized as a 'fitness landscape' with non-linear interactions such as lock-in, increasing returns,

Table 1.1 Overview of book

Chapter	*Contents*
PART 1 FOUNDATIONS	
1. Introduction	Introduction, including a brief historical overview of the development of systems thinking and recent developments in philosophy of science.
2. Philosophical foundations: Critical realism	An introduction to a range of philosophical issues and the way they are dealt with within critical realism.
3. Systems thinking and critical realism	This chapter demonstrates how systemic thinking and concepts lie at the heart of critical realism, and at the same time that critical realism can provide a philosophical underpinning for systems thinking.
PART 2 ONTOLOGICAL ISSUES	
4. Explanatory mechanisms	The idea of mechanisms or mechanistic explanation provides a clear link between critical realism (generative mechanisms), philosophy of science (explanatory mechanisms), and systems thinking (Maturana and mechanisms).
5. Emergence, generative powers and causality	From the previous chapter we can see that we need to develop more clearly concepts on the nature of systems, emergent properties, and causality.
6. Can social systems be autopoietic?	One of the major debates within systems thinking is the nature of social systems. Are there indeed systems that exist over and above the actions of individuals? If so, what is their ontological and epistemological status? This will be discussed through the vehicle of autopoiesis: self-producing systems.
PART 3 EPISTEMOLOGICAL ISSUES	
7. Observing systems: The question of boundaries	A discussion of boundaries and of developments in epistemology, especially the rise of soft systems and second-order cybernetics, in relation to critical realism and philosophy of science.
8. Knowledge and forms of truth	Developing a pluralist view of different forms of knowledge and considering their relation to truth.
PART 4. PRACTICE	
9. Methodology: What *might* we do?	The methodological implications that follow from the developments in ontology and epistemology outlined in the earlier chapters.
10. Ethics: What *should* we do?	The nature of knowledge and its link to action and hence to the place of ethics, focusing on the ethical stances of critical realism and Habermas's discourse ethics.
11. Conclusion	Conclusions and future issues.

punctuated equilibria and complex webs of interacting agents (Arthur 1994; Beinhocker 1997).

- Organizational strategy must change since the future is essentially unpredictable; markets do not attain equilibrium; and there may be sudden dramatic changes (note the credit crunch!) (Levy 1994; Stacey 2004).
- Within organizations there should be flat loose structures and networks of interacting, autonomous agents; periods of chaos should be expected, and patterns of behaviour may be 'attractors' (Lewin and Regine 1999; Murray 1998).

Certainly there seems to be much evidence in our globalized world that many of these effects are indeed real. However, with regard to the second question above, we would argue that all of the complexity effects can be generated within the traditional systems thinking framework as resulting from particular patterns of, especially, positive feedback loops and networks of interactions between large numbers of relatively simple units.

1.3 Structure of the book

The underlying structure of the thought behind this book can best be understood from Table 1.1. In essence, I am hoping that the book will draw out the commonalities between systems thinking, critical realism and philosophy more generally in a way that is of interest and value to all three domains. It should be noted that my own disciplinary areas are management science and information systems, and so when I use examples they will often be drawn from those areas.

Notes

1 One of the classic papers is titled 'What the frog's eye tells the frog's brain' (Lettvin *et al.* 1959).
2 Not to be confused with Maruyama's (1963) 'second cybernetics', which emphasized positive, reinforcing feedback loops as opposed to negative, balancing ones.
3 The paradigm idea was based originally on Thomas Kuhn's (1970) historical analysis of successive paradigms in natural science, but was developed into contemporaneous and competing paradigms in social science (Burrell and Morgan 1979).
4 For example, *iThink* from ISEE systems (www.iseesystems.com/).

2 Philosophical foundations

Critical realism

2.1 Introduction and context

Over the last 500 years, science has been incredibly successful in generating knowledge about the world and producing the vast array of technology that now shapes every minute of our lives. It is not surprising therefore that (natural) scientific knowledge came to be seen as the only valid form of knowledge. Modern-day science began in the days of the Enlightenment with the then novel idea that we could gain knowledge and understanding not from tradition or religious dogma or subjective contemplation but by observing and experimenting with the natural world itself. The origin of knowledge was to be *empirical* rather than metaphysical, a principle leading ultimately to the philosophies of empiricism and positivism that reached their apogee in the early twentieth century.

For empiricism, the source of all knowledge must be empirical – i.e. open to the senses and able to be observed by others. That which cannot be observed, directly or indirectly through instruments, ultimately cannot exist. The logical positivists (Schlick 1979) insisted that observations should be quantifiable and expressible mathematically, and that discussion of anything else was actually meaningless. Social science also strove to emulate natural science with the work of Compte (1986) and Durkheim (1951, orig. 1897) in sociology and Skinner's (1976) behaviourism in psychology.

However, empiricism and especially positivism were then subjected to major critiques by philosophers (e.g. Popper 1972), psychologists and sociologists, and virtually all their major tenets were discredited to some degree. Within natural science a degree of support was established around what was known as the hypothetico-deductive model, but social science became strongly split into what were seen as mutually incompatible paradigms (Burrell and Morgan 1979; Kuhn 1970). The main rupture was between functionalists who followed a broadly positivist approach and interpretivists who claimed that the meaning-based nature of the social world made it inherently unavailable to external observation and measurement. These developments were paralleled within systems thinking with the emergence of 'soft' systems and operational research, and second-order cybernetics (Checkland 1985; Eden 1993; Hayles 1999).

The pendulum swung fully away from positivism with the emergence of extreme anti-realist and anti-rationalist positions such as postmodernism (Rosenau 1992) and constructivism (Von Glasersfeld 1984). These, in various ways, denied the taken-for-granted assumptions of modernism: that there was an external world about which we could discover knowledge; that there were general standards of rationality such as true/false, good/bad or right/wrong; or that, especially in the social world, there was an underlying order or theory to be discovered rather than simply superficial surface happenings.

It is against this background that we can examine the development and contribution of critical realism (CR) as a significant philosophy of science able to underpin the systems disciplines.

2.2 Problems in the philosophy of natural science

In general, a *realist* understanding of science takes the view that certain types of entities – be they objects, forces, social structures, or ideas – exist in the world, largely independent of human beings; and that we can gain reliable, although not perfect, knowledge of them. However, as long ago as the eighteenth century Hume (1967, orig. 1750) and Berkeley (1995), for example, undermined such a view by denying fundamental tenets such as the existence of a physical world, causal necessity, or unobservable entities. Berkeley argued that we only actually know objects through our ideas and perceptions of them and that these, therefore, are all we can actually take to exist. Thus, to be is to be perceived. Hume was highly sceptical of several basic notions such as causality, unobservable entities, and induction. With regard to causality, he says we often see one event regularly followed by another and we believe that event A (e.g. swinging a bat) causes event B (a ball moving). However, all we can actually observe is the constant conjunction of the two events. Our belief that A *causes* B is simply that – a psychological belief. There is nothing more to causality than a regular succession of events. Hume is similarly sceptical about induction – the idea that witnessing an event occuring many times (e.g. the sun rising) warrants our claiming that it will always happen. These sceptical views, particularly those of Hume, underlie empiricism and have serious anti-realist implications.

During the twentieth century, 'naive realism' has been under constant attack from empiricism (which restricts science to mathematical formulations of empirical regularities) on the one hand and the many different forms of conventionalism or constructivism (that deny the existence of a world independent of human thought and perception) on the other.

2.2.1 Empiricism

In very broad terms, empiricism refers to those philosophies that see science as explaining events that can be empirically observed. That which is not manifest and capable of observation must be non-scientific or even, in the extreme view of the Vienna Circle philosophers, literally meaningless. Events are expected to display regularities or patterns that can be explained as being particular instances of universal laws of the form 'given certain conditions, whenever event X occurs then event Y will occur'. Science is seen as the systematic observation of event regularities, the description of these regularities in the form of universal laws, and the prediction of particular outcomes from the laws.

Logical empiricism was developed during the 1920s by a group known as the Vienna Circle, for example Schlick (1979) and Neurath (Neurath and MacGuinness 1987), who aimed to specify a truly scientific conception of knowledge and the world. Their main tenets were:

- Scientific knowledge must rest ultimately on that which is empirically open to the senses. This meant that any scientific propositions must be able to be empirically verified, and that anything unable to be directly or indirectly observed must be non-scientific or even meaningless.
- Empirical observations must then be reformulated into some strict mathematical or logical language (following the work of Frege, 1952, and Whitehead and Russell, 1925), generally expressed in terms of universal laws.
- There must be a unity of method across all sciences; thus social science and history must also be formulated in such a way.

These propositions rested on particular fundamental assumptions: (i) the idea that observation and perception were unproblematic – simply providing a mirror on nature; (ii) the Humean (1967, orig. 1750) principle that the observation of one event following another (e.g. one ball hitting another) did not enable us to prove some underlying causal mechanism – all that we can claim are 'constant conjunctions of events'; (iii) the principle of induction – that *universal* laws could be derived from a set of *particular* observations accompanied by the deduction of predictions from the laws.

This view of science was extensively critiqued. The idea of pure, objective perception and observation was exploded by psychologists (Gregory 1972; Piaget 1969), sociologists (Cicourel 1973) and philosophers (Hansen 1958; Merleau-Ponty 1962; Popper 1972). They showed, theoretically and experimentally, that the brain was not simply a blank slate on which the external world imposed itself, but rather that perception and conceptualization were active constructions of the nervous system. Hesse (1974), Popper (1972), Wittgenstein (1958), and Kuhn (1970) showed that observational terms, i.e. the language we use to describe our observations, were not an atomistic picturing of reality but part of a pre-given linguistic structure – in short that all observation was theory-dependent. And Popper (1959; 1969), based on Hume, rejected the possibility of induction and verification, replacing them with deduction and falsification.

In response to these criticisms there developed the 'deductive-nomological' (D-N) or 'hypothetico-deductive' method, centred around the work of Hempel (1965) and Popper. Science was still seen to be based fundamentally on empirical observations, although their theory-dependence was recognized. From such observations, theories were generated and expressed in terms of universal (nomological) laws ('covering laws'). Explanation, or prediction, then consisted of the logical deduction of particular events given some antecedent conditions and a set of laws. It was accepted that the laws might only be expressed in terms of statistical probabilities, and that they could not be *proved* to be true inductively. Some people maintained a confirmationist view that empirical evidence could provide support for a theory, while Popper developed the falsificationist approach that negative observations could definitely refute a theory. On this view, science should constantly aim to reject poor theories rather than support or confirm good theories. Hume's view of causation was still largely accepted. There was still general scepticism about the ontological status of theoretical concepts that could not be observed fairly directly, leading to debates about the legitimacy of 'theoretical entities'. Perceptibility was the criterion for existence.

The D-N approach also suffers from a range of problems, some of which will be explained in Section 2.2.2 on conventionalist alternatives. But, to highlight a few:

- Falsificationism, certainly in simple form, does not stand up – does a failed experiment falsify an underlying theory, or simply the experiment itself and its supplementary theories? Theories often need to be developed despite initial failures, not just abandoned (Lakatos 1978). Does not falsificationism implicitly rely on induction – i.e. moving from particular instances (of failure) to the general statement that it will always fail?
- The covering law model and especially Humean causality was impoverished, simply providing a description of *what* happened in highly constrained experimental conditions, with no explanation of *why* it happened, or sometimes did not happen, and with no mechanism for the generation of new theories or putatively real entities. This is particularly problematic from a realist point of view, as it restricts 'reality' to the domain of empirically observable events and prohibits underlying generative mechanisms.
- The D-N approach did not correspond, in many ways, with the actual practices of scientists and could not therefore satisfactorily explain the de facto success of science.
- The proposal that the social world was in essence no different from the natural world simply could not be sustained.

2.2.2 Conventionalism

Problems with the empiricist view of science centre on the impossibility of pure, unmediated observation of empirical 'facts'. So, the term conventionalism covers a wide range of philosophies that all emphasize the inevitable dependence of scientific theories on human perception, conceptualization and judgement: in short, conventions (Keat and Urry 1981).

The first position, *pragmatism*, derives from philosophers such as Dewey (1938) and Peirce (1878) and has been developed most radically (and perhaps somewhat illegitimately) by Rorty (1980; 1989). At a general level pragmatism is a view about the purpose of science – that it is essentially a practical activity aimed at producing useful knowledge rather than understanding the true nature of the world. Thus, Peirce developed a pragmatist theory of meaning such that the meaning of a concept was specified purely in terms of the actual practical effects that it would have. He also developed a consensus theory of truth as that which would come to be believed by a community of scientists in the long term, rather than as correspondence to reality (Habermas 1978). Dewey saw knowledge and truth as the outcome of processes that successfully resolved problematic situations.

The second position on the nature of science comes from those who study the actual practices of scientists and find that they do not correspond to the standard philosophical theories. This becomes more than mere description when it is used to critique the possibility of particular philosophical prescriptions. Kuhn's (1970; 1977) identification of major paradigms of thought throughout science is so well known as to need little exposition. The general idea is a development of the theory-dependence of observation – at any one time there is a broad, underlying theoretical conceptualization (e.g. Einsteinian physics) that is unquestioned within 'normal' scientific activity. This paradigm informs all actual experimentation, which is simply puzzle-solving within the paradigm. The failure of particular experiments does not refute, or even question, the basic paradigm. Only in periods of 'revolutionary' science, when there are many anomalies, do paradigms actually become questioned or compete.

This view leads to a much greater recognition of the social and psychological nature of scientific activity. A paradigm develops through consensus within a social community of scientists via many practical mechanisms such as learned societies, journals, or funding bodies. Individual scientists come to accept the underlying assumptions concerning research practice, theoretical validity, and core values as they become members of the community, often during their doctoral studies. Theoretical innovations that challenge the paradigm are generally rejected without serious consideration.

The basic idea of paradigms replacing each other over time has evolved, particularly within social science, to the idea of there being competing paradigms existent at the same time (e.g. positivist, interpretive and critical) (Burrell and Morgan 1979). This is often combined with the claim that paradigms are incommensurable (although Kuhn himself did not agree with this: Kuhn 1977). That is, each paradigm is so all-inclusive in defining its ontological and epistemological presuppositions that it is literally not possible to actually compare them – each defines its own 'reality'. Clearly the Kuhnian view has major relativistic implications for empiricism, since it points out the constructed, conventional nature of scientific theorizing and sees truth not as correspondence to some external reality but as that which is accepted by a scientific community at a particular point in time. The incommensurability thesis is even more undermining in that it makes it impossible to judge between paradigms or even to assert that a later paradigm is actually superior to an earlier one.

The third viewpoint, the sociology of scientific knowledge, can be seen as an intensification of Kuhn's study of the actual practice of science. It investigates the way in which scientific and technological knowledge comes to be constructed and accepted within a scientific community (Barnes 1977; Bijker *et al.* 1987; Bloor 1976; Collins 1985; Knorr-Cetina and Mulkay 1983; Latour 1987; Woolgar 1988). The most radical theories from this perspective (e.g. Bloor) argue that in fact science is no different from other forms of purposeful social activity and actually has no greater claim to truth.

2.2.3 The relationship between natural and social science

There have also been major debates over the nature of social science in relation to natural science that can only be sketched here (for overviews see Burrell and Morgan 1979; Giddens 1976; Keat and Urry 1981; Outhwaite 1987).

Broadly, there are three possible positions:

1 The *naturalist* view that there is one general approach to science that applies to all domains. Within this category, positivists hold that for anything to be scientific it must follow the canons of positivism/empiricism and thus be based on universal generalizations from empirical observations (Giddens 1974). This was in fact accepted by early sociologists such as Comte (1986) and, despite much criticism, continues in areas such as empirical and functionalist sociology and much management research. Critical realists, on the other hand, maintain a modified naturalism that is non-positivist and that accepts there are some differences between the natural and social worlds (Bhaskar 1979).

2 The antithesis is the view that the social world is intrinsically different from the natural world, being constituted through language and meaning, and thus involves entirely different hermeneutic (Bleicher 1980), phenomenological (Schutz 1972), or social constructivist (Gergen 1999) approaches. The argument here would be the idealist one that ontologically social objects do not exist in the way physical ones do (i.e. as subject independent), and that epistemologically there is no possibility of facts or observations that are independent of actors, cultures or social practices. The arguments of both Habermas (1978) and Giddens (1976) fall in this category.

3 The most radical position denies the possibility of objective or scientific knowledge at all, in either domain. Arguments here come from the strong sociology of knowledge programme discussed above; from post-structuralists such as Foucault (1980) who point out the extent to which even our most basic categories such as male/female are socially constructed, and the inevitable intertwining of knowledge and power; and more generally from postmodernists (Best and Kellner 1991) who attempt to undermine even the most basic categories of modernist rationality such as distinctions between truth and falsity and better or worse, or the existence of external reality.

2.3 An introduction to critical realism

Critical realism has been developing since the late 1970s (Bhaskar 1978; Bhaskar 1979; Bhaskar 1986; Bhaskar 1993; Keat and Urry 1981) in response to the fundamental difficulty of maintaining a realist position in the face of the criticisms, outlined above, of an empirical and naturalist view of science. Its original aims (which this chapter will concentrate on) were:

1 To re-establish a realist view of *being* in the ontological domain while accepting the relativism of knowledge as socially and historically conditioned in the epistemological domain (Bhaskar 1978). In other words, to establish that there is an independently existing world of objects and structures that are causally active, giving rise to the actual events that do and do not occur; but at the same time, to accept the criticisms of naive realism and to recognize that our observations and knowledge can never be pure and unmediated, but are relative to our time period and culture.
2 To argue for a critical naturalism in social science (Bhaskar 1979). That is, to maintain that the same general process of science is applicable in both the natural and social domains but to accept that the particular characteristics of the social world place inevitable limits on that process.

Originally Bhaskar referred to his work as either 'transcendental realism' or 'critical naturalism', reflecting these two aims, but these became contracted to 'critical realism'. In later work (Bhaskar 1993; Bhaskar 1994) the use of the qualifier 'critical' related also to critical social theory (Habermas 1974; Habermas 1978), and put forward the argument that no social theory can be purely descriptive: it must be evaluative, and thus there can be no split between facts and values. Following from this is the view that social theory is inevitably transformative, providing an explanatory critique that logically entails action (Archer *et al.* 1998, Part III).[1]

Critical realism is now becoming influential in a range of disciplines: geography (Cox 2013; Pratt 1995; Sayer 2013; Yeung 1997); economics (Fleetwood 1999; Lawson 1997; Lawson 2013; Spash 2012); organization theory (Ackroyd and Fleetwood 2000; Marsden 1993; Reed 1997; Reed 2001; Tsang and Kwan 1999); sociology (Archer 1995; Gorski 2013; Layder 1994; New 1995; Sayer 1997a); international relations (Fiaz 2013; Herborth 2012; Wright 1999); Marxism (Brown *et al.* 2002); psychiatry (Pilgrim 2013); medicine (Cruickshank 2012; DeForge and Shaw 2012); and research methods in general (Layder 1993; Miller and Tsang 2011; Sayer 1992; Wynn and Williams 2012).[2]

There have been significant developments following from the early work, and some of these will be discussed more fully in Chapters 3 and 10. In particular, we can mention the development of the ideas of emancipatory critique (Bhaskar 1986; Bhaskar 1991), the move to dialectical critical realism (Bhaskar 1993; Bhaskar 1994), meta-reality and the spiritual turn (Bhaskar 2000; Bhaskar 2002b; Bhaskar 2002c; Bhaskar 2002d), and applied critical realism and interdisciplinarity (Bhaskar 2010). The book *The Formation of Critical Realism* (Bhaskar and Hartwig 2010) is a personal reflection on the history of development of Bhaskar's thought, and Hartwig has edited a comprehensive *Dictionary of Critical Realism* (Hartwig 2007).

2.3.1 Arguments establishing an independent ontological domain

The first step is to put forward arguments that establish the existence of an ontological domain separate from the activities and cognitions of human beings.

Bhaskar's (Archer *et al.* 1998, p. 23) starting point is to argue, specifically in opposition to empiricism and positivism, that science is not just a matter of recording constant conjunctions of observable events but is about objects, entities and structures that exist (even though perhaps unobservable) and that generate, or give rise to, the events that we do observe. The form of the argument is a *transcendental* one (this follows a broadly Kantian interpretation of 'transcendental'). That is, it begins with some accepted happening or occurrence and asks: What must the world be like for this to occur or to be intelligible? In this case, what is accepted both by empiricism and by many forms of idealism is that we do have perceptual experience of the world, and that science is carried out through experimental activity in which scientists bring about particular outcomes.

The argument is that neither empiricism nor idealism can successfully explain these occurrences, and that they necessitate some form of realist ontology. With regard to perception, we can note that as human beings we have to learn (as babies) to perceive things and events; that our perceptions can change or be mistaken (e.g. visual illusions); and that scientists, for example, have to be trained to make observations correctly. These statements all imply that there must be a domain of events that are independent of our perceptions of them – what Bhaskar calls an *intransitive* domain – and, indeed, that these events would exist whether or not they were observed or whether or not there were even observers. Thus, there is a domain of actual events only a (small) subset of which are perceived and become empirical experiences. That which is not experienced is not known, but that does not mean to say that it does not exist. In other words, there is an infinity of events that do actually occur but that are never empirically observed.

Moving on to experimental activity, this shows several things. We can note that the experimenter causes (i.e. brings about) the experimental conditions but does not cause the results; these depend upon the underlying causal laws or mechanisms that are operative at the time. The regularities that are expected may or may not occur, and this depends partly on how well the experiment is carried out rather than on whether the presumed laws are or are not working. In fact, the occurrence of empirical regularities (i.e. constant conjunctions of events) in general is fairly rare – that is why the experiment is necessary to try to bring them about in the first place. The world is not full of constant conjunctions. But despite this, experimental results do in fact hold outside the experiment, as is attested by the enormous success of our technology.

The implications of this are that causal laws (more precisely, from a critical realist perspective, causal mechanisms) must be different from and independent of the patterns of events they generate; and that the experimenter aims to produce a constant conjunction of events by *closing* what would otherwise be an open system. Thus the intelligibility and success of experimental activity demonstrates the existence of an intransitive domain of casual laws separate from the events they generate. And the corrigibility of perception demonstrates the separation of events from particular experiences of them. This leads to a conceptual separation between a domain of causally operative structures or systems; the events that they generate; and those events that are empirically observed. Thus empiricism is doubly wrong in identifying causal laws with empirical regularities. It reduces underlying laws or mechanisms to actual events, and then reduces events in general to experiences.

The argument can be expressed in terms of the mistake that both empiricism and strong forms of idealism or conventionalism make – that is, the *epistemic fallacy*. The essential mistake is in reducing the ontological domain of existence to the epistemological domain of knowledge – statements about being (i.e. what exists) are translated into ones about *our* (*human*) *knowledge* or experience of being. For the empiricist, that which cannot be experienced cannot be. For the conventionalist, limitations of our *knowledge* of being are taken to be limitations on being itself. In contrast, the realist asserts the primacy of ontology – the world would exist whether or not humans did.

The argument so far establishes that given the successful occurrence of science, there must be an intransitive world of events and causal laws; but what exactly are causal laws? Or, rather, what is it that causes or generates events, given both the regularities that can be established in experiments and the common absence of regularity outside? Equally, how can we assure ourselves that event regularities are based on necessary connections rather than simply on coincidence? The answer is that there must be enduring entities, physical (e.g. atoms or organisms), social (e.g. the market or the family) or conceptual (e.g. categories or ideas) (Bhaskar 1997), observable or not, that have *powers* or *tendencies* to act in particular ways. The continual operation and interaction of these entities generates (i.e. causes), but is independent of, the flux of events.

'Entities' are structures, consisting of particular components that have certain properties or powers as a result of their structure. Thus gunpowder has the power to cause an explosion, a plane has the power to fly, a person has the power to compose music, a market has the power to generate wealth, and an inequitable distribution system the power to cause poverty. Entities may have powers without exercising them at a particular time (it may need an experiment or particular stimulus to trigger them), and powers may be exercised but not become manifest in events because of the countervailing operation of some other generative mechanism. The heart of this argument is that of a *causal* criterion for existence rather than a perceptual one. In other words: for an empiricist, only that which can be perceived can exist, whereas for a realist, having a causal effect on the world implies existence, regardless of perceptibility.

2.3.2 Critical realism and natural science

For Bhaskar, reality is both intransitive (existing independently of humans) and stratified – i.e. hierarchically ordered (Archer *et al.* 1998, p. 41). The first form of stratification is between structures or mechanisms; the events that they generate; and the subset of events that are actually experienced. These are known as the domains of the *Real*, the *Actual*, and the *Empirical* (see Figure 2.1). The Real contains mechanisms, events, and experiences – i.e. the whole of reality; the Actual consists of events that do (or do not) occur; and the Actual includes the Empirical, those events that are observed or experienced. These distinctions arise from the transcendental arguments above – namely that we should not reduce all events to only those that are observed, and we should not reduce enduring causal mechanisms to events.

A second form of stratification is within the realm of objects themselves (Archer *et al.* 1998, p. 66) where causal powers at one level (e.g. chemical reactions) can be seen as generated by those of a lower level (atomic valency). One stratum is emergent from another (what Bhaskar terms 'emergent powers materialism'). The picture of the real is thus one of a complex interaction between dynamic, open, stratified systems, both material and non-material, where particular structures give rise to certain causal powers, tendencies, or ways of acting often called by Bhaskar 'generative mechanisms' (Bhaskar 1979, p. 170). Although the term 'mechanism' sounds like a physical object, in fact Bhaskar uses the term to refer to the powers

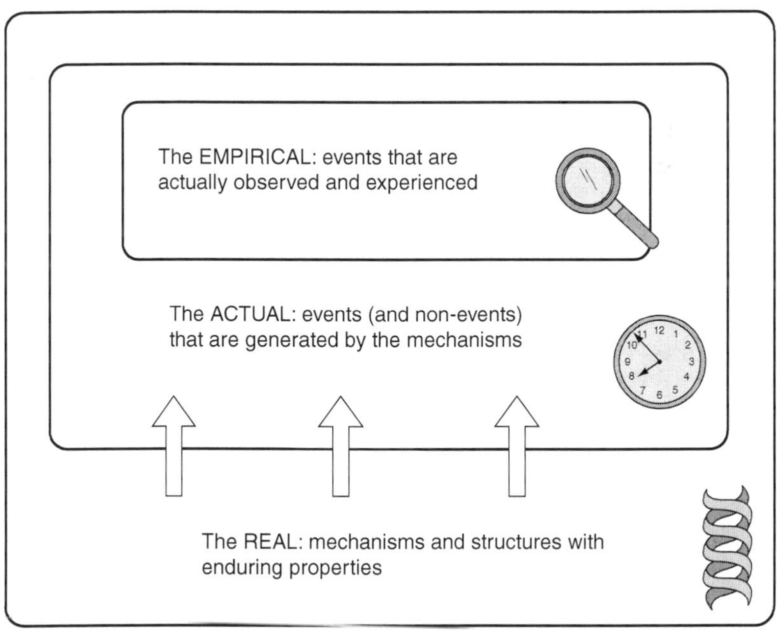

Figure 2.1 The Real, the Actual, the Empirical.

or properties of an object. For example, a plane has the generative mechanism of the power to fly. The interaction of these generative mechanisms, where one often counterbalances another, causes the presence or absence of actual events.

Having established the intransitive *objects* of knowledge, we must recognize that the *production* of knowledge is very much the work of humans, and occurs in what we could call the *transitive* dimension (Bhaskar 1989, p. 18). Acknowledging the work of sociologists, the practice of science is a social process drawing on existing theories, results, anomalies and conjectures (the transitive objects of knowledge) to generate improved knowledge of science's intransitive objects. This distinction allows us to admit the *epistemic* relativity of science, the fact that knowledge is always historically and socially located, without losing the ontological dimension. We should also note that such epistemic relativity does not imply a corresponding *judgmental* relativity, i.e. that all views are equally valid and that there are no rational grounds for choosing between them.

We can now characterize the realist method of science as one of *retroduction* (this is the same as 'abduction' as developed by Peirce – see Habermas 1978, p. 113 – in contrast to induction and deduction). We take some unexplained phenomenon that has been observed and propose hypothetical mechanisms that, *if they existed*, would generate or cause that which is to be explained. So, we move from experiences in the Empirical domain to possible structures in the Real domain. Such hypotheses do not of themselves prove that the mechanism exists, and we may have competing explanations in terms of other mechanisms, so the next step is to work towards eliminating some explanations and supporting others. Finally, this leads to a correction of previous results or theories. Bhaskar summarizes this as description, retroduction, elimination, identification and correction (DREIC) (Bhaskar 1994, p. 24).

An obvious objection is: how do we *know* that such hypothetical mechanisms actually do exist rather than being merely interesting ideas? At one level the answer is that we can never know for certain, since critical realism accepts that our knowledge is always ultimately fallible. More practically, however, the intransitivity of real structures means that they will always have the potential for effects that go beyond us, i.e. that are out of our control, and the methodology means that we should aim to eliminate alternative explanations by testing in some way for their potential effects.

So, the main feature of a critical realist approach to science is a fundamental concern for *explanation* in terms of independent underlying causal or generative mechanisms which may in principle be unobservable. This is in contrast to the empiricist approach, which limits itself to empirically measurable events and their abstraction into general laws; or the idealist approach, which has difficulty accepting a causally efficacious ontological domain.

2.3.3 Critical realism and social science

We now move to the second major argument of critical realism, that social science is essentially similar to natural science in its realist character albeit with modifications to reflect the particular nature of the social world. We can begin by asking, what would rule out a realist approach to social science? The answer is that a realist approach is ruled out if there are no intransitive objects for social science to investigate. Such an argument could come from the extreme constructivists (or superidealists as Bhaskar calls them), who would also apply it to the natural world; or from those, such as Checkland (1989), who would argue for the distinctive nature of social phenomena as being intrinsically meaningful and not existing independently of social actors. Space precludes a full discussion of this complex issue (see Archer 1995; Archer *et al.* 1998 Part III; Bhaskar 1979; Bhaskar 1994; Bhaskar 1997; King 1999a; New 1995; Outhwaite 1987), but I will outline i) the argument for intransitive social structures; ii) implications for the nature of societies; iii) the limits on naturalism that follow from i) and ii).

The primary argument (Bhaskar 1979, Chapter 2) is against methodological individualists, such as Popper (1962) (and Margaret Thatcher who claimed that 'society' does not exist!),who argue that all explanations can be couched in terms of an individual person's beliefs and actions. The first refutation concerns emergent properties. There are attributes of people that concern physical properties such as height or weight; there are attributes that we share with other animals such as pain or hunger; but there are many attributes, essentially human ones, that are unavoidably social, for example 'bachelor', 'banker', or 'nun'. These are only intelligible within the context of a social institution or practice (Searle 1996). The second argument is that many activities we undertake, most obviously perhaps language, must already exist and be available for people to learn and then use. As Wittgenstein (1958) argued, there can be no such thing as a private language – every time anyone has a conversation, uses a credit card, or waits for a train they are assuming the existence of a structured, intransitive domain of resources, concepts, practices, and relationships. The successful occurrence of social activities warrants the existence of causally efficacious, although unobservable, social structures.

Bhaskar (1979) does accept, however, that social phenomena are inherently different from material phenomena and that this does put limits on the nature of social science:

- Ontological
    - Social structures do not exist independently of the activities they govern, or, put another way, they exist only in their effects or occurrences. Social structures enable social activities and through those activities are themselves reproduced or transformed. Thus, they are themselves the result of social activity. In contrast, the laws of the natural world are not affected by their own operation.
    - Social structures do not exist independently of the agents' conceptions of what they are doing. Thus agency always requires some degree of interpretation and understanding of the meaning of the actions undertaken, although this does not imply that agents cannot be mistaken, and it does not require that they be fully aware of the consequences of their

activity. In contrast, natural phenomena are independent of our conceptions of them.
    - Social structures are localized in both space and time, unlike natural laws or tendencies that are generally universal. They hold only in particular cultures or sub-cultures and only for finite periods of time.

- Epistemological
    - Social systems are inherently interactive and open. While the same is true for natural systems, it is the case that they can be artificially closed or controlled in the laboratory, and this indeed is the principal reason for experiments. This however is not (generally) possible in social systems. The main effect is that it is difficult to test theories, since predicted effects may or may not occur depending on a multitude of factors. This point focuses attention on a theory's explanatory rather than predictive power.
    - The possibilities of measurement are very limited since intrinsically the phenomena are meaningful, and meanings cannot properly be measured and compared, only understood and described.

- Relational
    - Social science is itself a social practice and is, therefore, inherently self-referential. This means both that social science knowledge can itself affect the social world, and perhaps change it (e.g. the self-fulfilling prophesy); and that it is itself a social product and therefore will be shaped by the social conditions of its production. This does not make social science totally transitive – once an event has occurred, or some theory been produced, it becomes intransitive relative to possible explanations of it.
    - I would draw a second conclusion from this: that social theories must be self-consistent in not contradicting their own premises since they are part of their own domain.

All of the above place limits or constraints on the practice of social science, but do not make it different in principle from natural science. It is still driven by the existence of an intransitive domain of generative mechanisms; recognition of the epistemic (but not judgmental) relativity of knowledge; and a retroductive methodology that explains events by hypothesizing underlying causal mechanisms.

2.4 Criticisms of critical realism

Although this book clearly propounds a critical realist philosophy, we should still consider criticisms of and objections to critical realism. In fact, it is interesting that little has been written as a direct critique of CR, especially within the philosophical literature. We may speculate that this is partly due to Bhaskar's disengagement from the philosophical establishment – he has never had a significant academic position, always remaining independent; he writes books but rarely papers and so is not well established in the mainstream journals; and he does not really engage in philosophical conferences and debates. His work has mainly been picked up in other disciplines, especially the social sciences, where the reception has usually been positive rather than critical. Indeed, even some of the critics discussed below (e.g. Chalmers and Callinicos) end up saying that despite their concern with particular arguments, they basically think CR is correct!

The first point of debate is the status of one of the main planks of CR – the transcendental argument for an independent, stratified ontological domain. This form of argument is the reverse of the traditional syllogism – it goes from the agreed occurrence of some phenomena (in this case scientific experimental activity and the exportation of its results to the real world) backward to an inference about what, therefore, the world must necessarily be like (independent stratified ontology with causal powers):

> The intelligibility of experimental activity presupposes then the intransitivity and structured character of the objects of scientific knowledge, at least in so far as these are causal laws. And this presupposes in turn the possibility of a non-human world … and in particular of a non-empirical world.
>
> (Archer *et al.* 1998, p. 26)

Doubt can be cast on the strength of this argument in several ways. It seems to rest very much on what is meant by 'intelligible'. If it simply means understandable or explainable, then this seems quite a weak argument. Does intelligibility really imply the existence of an external world, or does it just imply that scientists have that belief, whether or not it is actually true? We could similarly argue that the intelligibility of religious activity implies the existence of God, but presumably we would only wish to argue that it implies a *belief* in God on the part of religious people. The argument also relies on the belief that there is only one possible way of accounting for the activities of scientists. But if there is, in fact, another possibility then does not the argument fall down? Clarke (2010) suggests that a similar but not identical transcendental argument is that of Cartwright (1999).

In fact, does not the argument rest on the *success* of science rather than on its intelligibility (Chalmers 1988)? In other words, it is not so much what scientists believe about what they are doing, but the fact that knowledge generated through experimental activity is found to hold outside the experimental situation as testified by the enormous developments of successful technology. This would be the viewpoint of *scientific realists* such as Putnam (1975) or Chakravartty (2007), who do not see the need for the transcendental argument.

We might also question whether the premises about experimental activity are actually shared by competing positions or, indeed, are an indubitable description of science anyway (Callinicos 1995). How do we know that there are not competing theories about scientific practice and that these do not offer different accounts that still make the activity intelligible? Here Bhaskar would probably argue that his is an *immanent critique.* That is, his arguments are always contextual and directed against particular positions, in this case empiricism and some forms of idealism, rather than being totally general. There may well be other views on the nature of experiments, e.g. from a postmodern perspective, but then the nature of the argument would be different. McWherter (2012) has developed a systematic reconstruction of Bhaskar's transcendental argument which addresses several of these points.

Finally, we could object that even if we accept the premises, the nature of the conclusions depends very much on the general scientific knowledge of the day. If a Greek or medieval philosopher attempted a similar argument they would come up with a very different picture of the nature of the world. I think this argument has to be accepted but is compatible with CR's wider acceptance of fallibility. Bhaskar accepts that knowledge is temporally relative and will change, and even accepts that CR itself is only 'the best explanation so far' – 'the transcendental consideration is not deployed in a philosophical vacuum: it is designed to situate, or replace, an existing theory; and may of course come, in time, to suffer a similar fate' (Bhaskar 1979, p. 6).

A second area of concern is the extent to which the theory of science is simply descriptive or is actually normative and, if so, what are the strengths of its prescriptions. Many would agree that CR (Baert 1996), with its acceptance of unobservable entities, the role of metaphor and analogy, and the importance of explanation, is a much better description of the activities of actual (natural) scientists than empiricism or even Popperianism. To what extent, however, does it provide powerful normative procedures for natural science; and to what extent does it apply to the activities of social scientists (Cruickshank 2007)?

Methodologically, the description, retroduction, elimination and identification formulation has several weaknesses. Given the acceptance of the subjectivity of the transitive domain and the theory dependence of observations, it seems unlikely that one can begin with objective and agreed descriptions of particular phenomena. Any description will already be imbued with underlying theoretical concepts and in the social sciences will also be highly value-laden.[3] This will clearly condition the forms of generative mechanisms that are postulated to explain a phenomenon and will make any sort of comparison or contrast very difficult.

Retroduction itself is an intuitive and creative process, rather than a logical one.[4] This is a necessary part of scientific endeavour, but it can result in a proliferation of possible explanations, some of which may well be untestable, or at least unrefutable. This places a lot of weight on the latter stages of elimination and identification, but here CR runs into problems because of its critique of traditional empirical testing, verification, and induction. How is the scientist, especially the social scientist, ever going to be able to undertake testing that unambiguously rules out or rules in particular hypothetical mechanisms, particularly when such mechanisms may be unobservable, and their powers may be unactualized?

This is related to a third problem: the nature of truth within critical realism. While the basic orientation is towards a correspondence theory of truth, i.e. that knowledge in the transitive domain in some sense corresponds to its objects in the intransitive domain, the acceptance of epistemic relativity means that we can never prove or be certain that this is the case. This potentially brings in elements of a consensus theory of truth. Bhaskar himself recognizes four dimensions of truth (Bhaskar 1994): *normative-fiduciary*, truth as that which is believed by a trustworthy source; *adequating*, truth as based on evidence and justification rather than mere belief; *referential-expressive*, truth as corresponding to or at least being adequate to some intransitive object of knowledge; and onto*logical/alethic*, the truth of things in themselves and their generative causes in the intransitive domain, i.e. no longer tied to language although expressible in language. The fourth aspect is clearly controversial (Groff 2000). We are thus left with a problem of precisely what criteria we can use to judge between competing explanations if not a clear view of truth.

A fourth area of criticism concerns that of naturalism – i.e. the extent to which an approach developed largely in relation to natural science can be applied to social science. Clearly Bhaskar recognizes the fundamentally different nature of the social world and the limitations this places on science. But are not these limitations in fact so great that CR-type science is not possible? Giddens (1976) recognizes that even natural science involves a transitive, hermeneutic domain but that social science involves a double hermeneutic in that the objects of knowledge are themselves intrinsically socially structured and human-dependent. If social 'structures' are unobservable, and indeed only exist through peoples' activity; if social systems are open and not amenable to experiment; and if social activities always rely to some extent on prior common sense or theoretical conceptualization – then to what extent is it really possible to test competing explanations and identify 'true' ones?

Coming from the opposite direction, King (1999b) argues against the realist notion of a causally effective social structure over and above the knowledgeable actions of individual agents. He suggests that Bhaskar's concept of social structure involves two contradictions (or 'antinomies'). The first is that society is both dependent on individuals and also independent of individuals. From Bhaskar's viewpoint this apparent contradiction is resolved through the idea of emergence. Society, as a separate ontological entity, emerges from but is separate to the activities of individuals. This allows for the development of a social theory with two separate types of entity – individuals and society – that interact with and mutually shape each other. King objects that such a view of society is a reification and that in fact:

> The apparently structural and emergent aspects of society can be successfully accounted for by hermeneutic reference to individuals and their meaningful interactions with other individuals alone.... Social reality is coextensive with the individuals involved and is neither more nor less than those individuals.
>
> (King 1999b, pp. 271–272)

The second antinomy is that social action is said to be always intentional, yet is also said to be non-intentional and materially caused. The point at issue is related to the previous one – to what extent should individual action be explained in terms of external social and material structures as opposed to simply the intentions of the individual? This is clearly a major debate within social theory and there will be further discussion in Chapter 6.

The fifth area of debate I will discuss is the nature and extent of critical realism's claim to be 'critical' not so much in the epistemological sense but in the political sense of bringing about change in society.[5] The idea is that social science is not value-neutral description but is inevitably explanatory critique of the status quo.[6] Social science concepts must always be evaluative or moralized, never purely descriptive. For instance, it is more correct to say 'Two children were murdered' than 'Two young humans ceased functioning', since it is a more precise and accurate description requiring a more specific explanation. Social science will always reveal examples of false beliefs, unmet needs and unnecessary suffering; and will often be able to identify their structural causes. Other things being equal, it is then possible to condemn the causes and propose action to remove or absent them. We thus move from fact to values and from values to actions in support of a transformation of society.

Sayer (1997b) accepts these arguments at a general level but points out the difficulty of enacting them in practice. In particular, it is not difficult to find many examples of false beliefs or suffering, but doing something about them requires both a correct identification of their causes and specific changes that are both desirable and feasible and that do not generate new problems elsewhere. The world is now highly complex and incredibly interdependent. Particular events or problems will often have multiple interlocking structural causes which are very difficult to untangle, and possible changes will often have undesirable and unintended consequences and will have to contend with an increasing diversity of values and cultures.

Baert (1996) maintains that Bhaskar's social theory is actually much better at explaining why societies remain the same than explaining why they are transformed. Certainly it is true that Bhaskar's transformational model of social activity (TMSA) emphasizes the way in which social actors necessarily draw on an already existing social structure and through their interactions reproduce it, and only potentially transform it.[7] Archer (1990; 1996; 1998) has addressed this point to some extent in her morphogenetic model, which emphasizes the independence of society from individual actors and therefore allows both reproduction and transformation through their mutual interaction. Baert also suggests that the TMSA model undervalues the extent to which social actors (not just social scientists) can develop their own discursive, theoretical knowledge of society and act on it to change rather than merely reproduce social structure.

Fine (2002) is particularly concerned with economics, where CR has generated a significant attack on traditional theory, especially econometrics (Fleetwood 1999, 2001, 2002; Lawson 1997, 2013). Interestingly, rather than being a supporter of the status quo (in economics), Fine argues that CR is neither critical nor realist enough to have much effect. It is not critical enough because it has largely confined itself to critique at the level of methodology rather than substantive theory. Fine suggest that mainstream economists (and perhaps this can be extended to other disciplines) have no interest in methodology, or indeed realism or the real world. CR is not realist enough in not having significant theoretical conceptions of core economic phenomena such as capital and capitalism. One could reply that Bhaskar has always maintained that the *philosophy* of CR is intended as a foundation for specific sciences, not as a replacement. So now perhaps is the time for critical realists within the disciplines to use it to generate more and better substantive theories and to prove its worth in practice.

2.5 Conclusions

This chapter has set out the groundwork for the rest of the book by introducing the basic, and original, claims of critical realism. It began by looking at the historical context that led to the development of critical realism in the first place and the later context that has contributed to its recent rise in popularity. It then described the most fundamental arguments of CR as they were developed in the original books by Bhaskar – *A Realist Theory of Science* and *The Possibility of Naturalism*. This chapter has not discussed the later developments such as the move to a dialectical version of CR, or its social theory. These will be discussed in more detail in Chapter 3, where we identify the links to systems thinking; Chapter 4, where we consider the core notion of mechanisms; Chapter 5, where we discuss the idea of causal powers or dispositions; Chapter 6, where we consider social systems and particularly the link to autopoiesis; Chapter 9, where we review the methodological implications of CR; and finally Chapter 10, where we look at its critical and ethical concerns.

In the final section above we discussed a variety of criticisms that have been raised, mainly from a philosophical perspective. It is interesting that many of those who put forward the criticisms are nevertheless generally supportive of CR, and the general view is that these are blemishes that can be dealt with, rather than fatal flaws. I hope that the rest of the book will demonstrate the huge promise that CR holds out across a broad range of practical, social and philosophical questions and problems.

Notes

1 The later work, especially on emancipation and dialectical critical realism, will be discussed further in this volume, especially in Chapter 10.
2 For two special issues devoted to critical realism in business and management, see *Organization Studies* (32:7), 2011, and *MIS Quarterly* (37:3), 2013. For one covering a wide range of fields see *New Formations* (56:Autumn), 2005.
3 A point Bhaskar clearly accepts.
4 Indeed Peirce, who coined the term, called it basically guesswork.
5 This is discussed more extensively in Chapter 10.
6 This is in direct opposition to positivism's insistence on a separation between fact and values.
7 There are indeed many similarities with Giddens' theory of structuration, which is also criticized as being overly regulative.

3 Systems thinking and critical realism

3.1 Introduction and context

Critical realism (CR) clearly embodies systemic and holistic themes at its very heart, with concepts such as totality, holistic causality, emergence, open systems, autopoiesis, and levels of stratification. These concepts have their own history of development within the discipline known as systems theory or systems thinking, but there is almost no reference to this literature within Bhaskar's (or other critical realists') writings. I do not know whether this reflects a lack of familiarity with the literature or a desire to establish a new and autonomous discourse which is not seen as a part of something else. In any case, the links are in fact very clear and it is the purpose of this chapter to draw them out.

I believe this is important to do for several reasons. First, it should be done for purely scholarly reasons. Systems thinking has a long history and its implicit contribution to critical realism should be made clear. Second, and more importantly, CR can gain from such an interchange. There has been much debate and clarification about these concepts within the systems literature which can aid their employment within CR, and there are further concepts and perspectives which CR could usefully employ. Third, systems thinking can, in its turn, gain from CR. Philosophically, it is still based on a schism between positivism and interpretivism which CR does much to dissolve.

The logic of this chapter is that the first section will present a brief history of the development of systems thinking since the 1920s, not for its own sake but in order to explore the various theories, concepts and debates that are relevant to CR. The next two sections then discuss the use of systems concepts in Bhaskar's early work (before *Dialectic: The Pulse of Freedom*, 1993) and later work.

3.2 The development of basic systems concepts

Systems thinking or the systems approach[1] developed in its modern form with a burst of new ideas in a range of disciplines during the 1920s and 1930s, although some of the underlying principles can be traced back to the Greeks, especially

Aristotle.[2] Traditional disciplines that were involved include biology, psychology and even quantum physics, while new disciplines emerged, such as ecology and cybernetics, based on systemic ideas.

During the 1970s there was a major epistemological break within systems thinking in which a new stream of thought based on constructivism or phenomenology was initiated. This mirrored similar developments within the other social sciences, which are obviously very much the concern of critical realism. This development is generally known as *soft* as opposed to *hard* systems thinking, or sometimes *second-order* rather than *first-order* cybernetics.

3.2.1 Phase 1, 'hard' or first-order systems thinking

The most fundamental idea of systems thinking is the anti-reductionist one that we cannot explain the behaviour of objects and entities purely in terms of the nature and constitution of their *parts* or *components*. Rather, the parts are related together in such a way that the whole has behaviours or, more generally, properties that are distinct from, and irreducible to, the properties of the parts. This is often expressed in the phrase, possibly due to Aristotle, that the whole is more than the sum of its parts. This is easily shown by examples: water has very different properties from its constituents oxygen and hydrogen; a plane can fly, its parts cannot; stereograms and 'magic eye' pictures can generate 3-D images in a human brain; and a football crowd can produce a 'Mexican wave'.

This may seem obvious now, but in the early part of the twentieth century it was totally against the prevailing worldview. Science had been incredibly successful over several hundred years based on the Cartesian reductionist view that the way to proceed was to successively split up entities into their component parts until ultimate components were reached, at which point ultimate explanations were possible. However, at this time Kuhnian-type problems were being experienced in all the major disciplines, even physics itself, and this led to the recognition of the importance of *wholes* over parts, or equally *form* over substance.

In biology, great progress had been made in understanding the parts of organisms, down to the level of their biochemistry, but this could not explain the complex behaviour of cells as a whole, nor could it explain how cells differentiated during the development of an organism. Two alternatives to reductionism emerged – vitalism and organicism. Vitalism (Driesch 1908) asserted that there must be some unknown or unobservable element or force that was possessed by living things, while organicism (Ritter 1919) held that the explanation was simply the organization of the relationships and the interaction of all the parts together. Early organicists actually used the term 'system' and it was perhaps best articulated in Woodger's *Biological Principles* (Woodger 1929).

Similar ideas were being developed in other disciplines. In psychology, the *Gestalt* school argued that perceptions and thoughts always occurred as wholes in themselves, which could not be broken up into parts (Ritter 1919). This is often illustrated by perceptual illusions such as the young/old lady, where we see one thing or the other but never both. Ecology was also picking up on ideas of relationships and wholeness from the organicist biologists. Haeckel (1866) created the term as 'the science of relations between the organism and the surrounding outer world', and von Uexküll (1909) named the outer world *Umwelt* or 'environment' – another key systems concept.

Finally we can mention that atomic physics itself, the bastion of reductionism, also recognized wholeness at the very fundamental levels of subatomic particles which are not so much discrete particles but webs of interacting forces. As Heisenberg (1963, p. 107) put it:

> in modern physics the world is not divided into different groups of objects but rather into different groups of relationships.... The world thus appears as a complicated tissue of events, in which connections of different kinds alternate or overlap or combine and thereby determine the texture of the whole.

A second important realization came out of quantum physics, again in opposition to the prevailing positivist view of science: the inevitable involvement of the *observer* in any observations or descriptions that we make of the world. Heisenberg's uncertainty principle showed that the results we might get could not be simply reflections of the external world alone but were always in part due to the very act of observation. Again, as Heisenberg said, 'Natural science does not simply describe and explain nature ... it describes nature as exposed to our method of questioning' (1963, p. 75). As we shall see, it is very much one of the important planks of systems thinking that the observer must be recognized as part of the system. The central systemic idea – that the characteristics and behaviour of entities depended on the structure of *relationships* between components rather than the properties of the components themselves – carries with it several other concepts – *emergence*, *hierarchy* (or stratification as Bhaskar tends to call it) and *boundaries*.

Emergence is certainly a key feature of Bhaskar's critical realism, which he has at times described as 'synchronic, emergent powers materialism'. Although the subject of much debate (Elder-Vass 2005), the basic idea of emergent powers or properties is clear. The emergent properties of an entity are properties possessed only by the entity as a whole, not by any of its components or by the simple aggregation of the components (as for example with mass). Emergent properties result from the properties of components and the particular structure of relationships between the components which constitute the entity. The examples presented above are all illustrations of emergent properties. This will be discussed more fully in Chapter 5.

With emergence comes hierarchy. If we consider a system at a particular level it consists of components and relations. However, each component can itself be treated as a system and 'opened up' to reveal another set of components and relations. This process can in principle go on for an indefinite number of levels until we reach the bedrock of indissoluble forces. We can also go in the other direction from the initial system and see that it is only a component of a further hierarchy of wider systems. In fact the term hierarchy can be misleading – it is better described as a nesting of systems within systems, much like Russian dolls. At each level, systems, with their emergent properties, interact with each other in a manner governed by their structure of relationships, generating a new level of system with its own emergent properties.

The third concept is that of boundary. If emergent properties are attributed to a particular entity in virtue of its components and relations, we must be able to demarcate the system that has these properties from its environment. This may seem relatively clear when we are dealing with physically discrete objects that have a single clear boundary. It becomes much more contentious when dealing with complex systems that may be physically diffuse or that may consist of different types of components, some of which may not actually be physical (e.g. information or ideas), and above all when we deal with social systems. Boundaries will be discussed more fully in Chapter 7.

The concepts covered so far may be considered structural in that they deal with the *structure* of systems as opposed to their *processes*. The distinction between these is time-relative, but essentially the structure of a system comprises the components and relations between components that remain (relatively) constant over time. Process or dynamics is that which changes. The main researcher in this area, who is often seen as the founder of the systems movement, is Ludwig von Bertalanffy (1950; 1971) (published in German in the 1940s) with the concept of *open systems* and of *general systems theory*.[3] Until that point, science had generally concerned itself with systems that were closed to their environment. Within such systems the second law of thermodynamics held, and this suggested that entropy would always increase and that the systems, including the whole universe, would eventually run down. However, this was clearly not the case with individual organisms or with evolution as a whole, both of which appeared to be anti-entropic. To resolve this dilemma, Bertalanffy proposed the concept of an open system which was in a state of dynamic (rather than static) equilibrium based on continual import from and export to the environment. Metabolism in the cell was one of the classic examples of this, and the *self-regulation* of these processes was one of the key emergent properties.[4] Another concept developed at this time, based on principles of feedback, was that of *homeostasis* (Cannon 1939).

Bertalanffy's second contribution was to try to establish a new, over-arching discipline known as general systems theory. This was based on the recognition that the systems concepts and principles we have described can be applied irrespective of the particular nature or substance of the systems concerned. It is therefore possible to study systems, relationships and organizations in the abstract and then apply them, as with mathematics, to particular domains.

The next, and very significant, development happened during and after World War II with the development of an entirely new discipline – *cybernetics*, the science of communication and control.[5] The early cyberneticians, Weiner (1948), von Neumann (1958), Shannon (Shannon and Weaver 1949) and McCulloch, were mainly mathematicians and engineers who were interested in the ways in which systems, both mechanical and biological, regulated and controlled themselves in a largely automatic way. They recognized that the key to this was in the concepts of *information* and *feedback*. Used initially in the design of self-controlling weapons, the ideas soon spread into modelling the functioning of the brain (McCullough and Pitts 1943) and developing the first digital computers (von Neumann 1958), as well as into anthropology (Bateson 1973b) and psychiatry (Bateson 1973b).

The most fundamental idea of cybernetics is that of *circular* (as opposed to linear) *causality*, more commonly known as *feedback*. This is a most ubiquitous phenomenon in which a chain of causal connections is such that a change in one element eventually feeds back to either balance or reinforce the initial change. This had been known about in practical terms for centuries. Philon of Byzantium (third century BC) designed an oil lamp which maintained a constant level of oil by means of a float, and Watt's steam engine governor was one of the most important inventions of the industrial revolution. Watt's governor involved two heavy metal balls connected to the output axle of the steam engine. When the axle speeds up, the balls move outward under centrifugal force; they are connected to the steam control so that the outward movement reduces the amount of steam, which in turn reduces the speed of the axle. If the speed reduces, the balls move inwards with the opposite effect. This is a classic example of negative or balancing feedback, which automatically maintains some variable at a constant level. Much of the regularity and constancy of the natural world is maintained by complex feedback loops such as this.

The opposite type of feedback also exists: positive or reinforcing feedback. Consider compound interest: a sum of money generates interest, which is added to the money so that even more interest is produced in the next period. While negative feedback produces order and stability, positive feedback produces exponential growth or decay. The early cyberneticians were mainly interested in the way that negative feedback could produce apparently purposive or teleological behaviour without any form of conscious mental control (Rosenblueth *et al.* 1943), although later Maruyama (1963) focussed attention on positive feedback processes, in what he called 'second cybernetics'.[6]

Systems concepts were also applied extensively in sociology: for example, by Parsons (1951), whose work was criticized for being overly functionalist; by Buckley (1967), who emphasized the dynamic and processual aspects of systems; by Luhmann (1982b), who produced a radical reworking of Parsons based on autopoiesis; and by Habermas (1978), whose work formed the basis for critical systems thinking (Jackson 1991; Mingers 1992b).

The final development I will consider in the first phase of systems thinking is what became known as system dynamics. Jay Forrester, at MIT, was initially interested in applying the ideas of positive and negative feedback to investigating population dynamics – especially patterns of urban development (Forrester 1969). He also applied the ideas to industrial supply networks (Forrester 1961). Perhaps the most famous analysis was the Club of Rome's report on the future of the world economy called 'Limits to Growth' (Meadows *et al.* 1972), which was one of the first publications to point out the effects of the world using up its natural resources.

However, from a critical realist perspective it is the later work of Sterman (2000) and Senge (1990) which is most relevant. They drew the distinction between the overt behaviour of the system of interest and the underlying, and often unobservable, pattern of causal relations that generated the behaviour. Their motto was that 'behaviour follows structure': if different people are put within the same structure, it is likely that the same behaviour will emerge. This is very similar to Bhaskar's distinction between the domain of actual events and the domain of the real enduring mechanisms which generate them (Mingers 2000a). Senge also developed the concept of 'systems archetypes', that is, particular patterns of feedback loops that occur very often in the real world and generate particular patterns of behaviour.

3.2.2 Phase 2, 'soft' or second-order systems thinking

The work so far described was carried out within the prevailing positivist paradigm but, as in other disciplines, this was extensively critiqued during the 1970s and a new paradigm, known as soft systems or second-order cybernetics, emerged. Within cybernetics there had always been a recognition that observation was not wholly objective, but was to some extent dependent on the act of observation, or the observer. Heinz von Foerster and others at the Biological Computer Laboratory saw that in using cybernetic ideas to study the mind and the brain they were in fact also studying the process of observation itself (Lettvin *et al.* 1959; Von Foerster 1984). This self-referentiality was referred to as the 'cybernetics of cybernetics' (Von Foerster 1975).

The ideas were developed most coherently in the work of the biologists Maturana and Varela (Maturana and Varela 1980, 1987; Mingers 1995b) who coined the term *autopoiesis* to described the circular, self-producing, organization of living systems, a term that has been explicitly used by Bhaskar (Mingers 2004a). Maturana wanted to answer the most basic question: what distinguishes living systems from non-living systems? What is their essential property? He saw that you could not characterize a living system, for example an amoeba, in terms of purpose, for it has no purpose except its continued existence; you could, however, characterize it in terms of what it *does*, what it produces. A living cell is a complex network of processes of chemical production that produces the very components which constitute the network in the first place. It produces itself. Non-living systems, which he termed *allopoietic*, produce something other than themselves: for instance, a chemical reaction converts some inputs into a different output. Autopoietic systems are organizationally closed but interactively open.

Maturana was also a neurophysiologist and had conducted empirical work on the perceptual systems of animals such as pigeons and frogs (Maturana *et al.* 1960, 1968). This work showed that there was not a one-to-one correspondence between the visual environment and the resultant neuronal activity. Perception could not be an internal picturing of the external world but was, rather, the result of internal patterns of correlation and association. Sensory stimuli did not determine but only triggered or selected subsequent states of nervous activity. The nervous system, too, is organizationally closed – it is not open to the environment; our experiences are internally constructed, although modulated through our interactions with the external world. This led to a strongly constructivist view of epistemology and ontology. The world that we experience, whether perceptually or linguistically, is a world that we construct; we can never have unmediated access to an external world:

> Indeed, everything said is said by an observer to another observer that could be him- or herself.
>
> (Maturana 1988, p. 27)

> I am saying that all phenomena … are cognitive phenomena that arise in observing as the observer operates in language.… Nothing precedes its distinction; existence in any domain, even the existence of the observer themselves, is constituted in the distinctions of the observer.
>
> (Maturana 1988, p. 79)

The second major source of interpretive thinking was within applied systems, specifically within engineering and management systems. Hard systems thinking had developed within engineering, for example in designing complex chemical plants. It had also developed within management as, for example, with Stafford Beer's management cybernetics (Beer 1966) or Ackoff and Emery's 'purposeful systems' (Ackoff and Emery 1972). But a new paradigm was established with Checkland's development of soft systems methodology (SSM) (Checkland 1972; Checkland 1981; Checkland and Scholes 1990; Mingers 2000b).

On the basis of many practical projects in organizations, Checkland argued that social systems were intrinsically different from physical systems. One could not take the nature of a social system as given, from an external viewpoint, in the way that one could perhaps a machine or an organism. The essential difference is that the members of a social system, such as an organization, would inevitably bestow their own meanings and senses on the system, and these had to be seen as equally valid ways of interpreting their reality. The purpose of SSM was, therefore, not to describe or design some objective system, but instead to articulate and explore the differing perceptions or worldviews held by participants within a problematic situation, and by doing so possibly to bring about an agreed improvement to the situation.

> [We] need to remind ourselves that we have no access to what the world is, to ontology, only to descriptions of the world, … that is to say, to epistemology.… Thus systems thinking is only an epistemology, a particular way of describing the world. It does not tell us what the world is. Hence, strictly

> speaking, we should never say of something in the world: 'It is a system', only: 'It may be described as a system'.... The important feature of paradigm II [soft systems] as compared with paradigm I [hard systems] is that it transfers systemicity from the world to the process of enquiry into the world.
>
> (Checkland 1983, p. 671)

Checkland explicitly allied SSM to phenomenology and against positivism, and he has never accepted the possibility of an excluded middle – namely critical realism.

3.2.3 Critical systems thinking

Finally, I should mention that, like other social sciences, systems thinking has also developed a critical stream, drawing mainly on the work of Habermas (Flood and Jackson 1991; Jackson 1985; Midgley 1995; Mingers 1980; Mingers 1992b). This recognized the role and limitations of both hard and soft systems thinking, and maintained that there was also a need for emancipatory systems thinking. And there has even been work drawing on postmodern perspectives and particularly on Foucault's work (Brocklesby and Cummings 1996; White 1994; White and Taket 1996). Recently, consideration has been given to ethics: Mingers (2009) undertook a comparison between Habermas's discourse ethics (Habermas 1992a) and the ethics implicit in critical realism (see Chapter 10).

3.2.4 Non-linear dynamical systems (complexity theory)

Complexity theory, also known as non-linear dynamical systems theory, developed during the 1970 and 1980s in a range of sciences – biology, chemistry, mathematics and economics (Kaufmann 1995; Waldrop 1992). Traditionally, these hard sciences had assumed stability, equilibrium, linear change, cyclicality, robustness, and simple models generating simple behaviour (and vice versa). Chaos and complexity are the results of a Kuhnian revolution that emphasizes instability, far-from-equilibrium, sudden change, sensitivity to initial conditions, and complex behaviour from simple models (and vice versa) (Lewin 1992; Mainzer 1997). Two interesting questions are: to what extent do these insights apply to soft sciences and organizations (Byrne 1998; Cilliers 2000)? And, to what extent can complexity theory be encompassed within traditional systems thinking?

Certainly there seems to be much evidence in our globalized world that many of these effects are indeed real at a social and economic level. However, with regard to the second question we would argue that all of the complexity effects can be generated within the traditional systems thinking framework as resulting from particular patterns of, especially positive, feedback loops and networks of interactions between large numbers of relatively simple units. For instance, Mosekilde and Laugesen (2007) have shown that the Beer Game, a well-known feedback-based management game, can display all the behaviour typical of complex systems.

3.3 Systemic concepts in Bhaskar's early work

In considering systemic motifs in Bhaskar's work, I shall distinguish between the early works, where they are relatively implicit, and *Dialectic: The Pulse of Freedom* (Bhaskar 1993) and *Plato Etc.* (Bhaskar 1994), where they become much more explicit but where different terms are often used.

3.3.1 Systems, structures, mechanisms and emergence

The Introduction of *A Realist Theory of Science* (Bhaskar 1978) outlines the fundamental concepts from which the initial version of critical realism (CR) is built. The world is taken (on the basis of transcendental arguments) to consist of *structures* and *mechanisms* (or 'things', although that term has overly physicalist overtones for his later work) that have *powers* and *liabilities* to generate the events that actually occur. These structures are distinct from the events they generate. Events occur at a particular point in time, but the structures are relatively enduring, exercising or not exercising their causal powers in interaction with each other.

The distinction is recognized between *closed* and *open systems* (and he does use the term 'system' here), where the former allow constant conjunctions of events, the Humean version of causality, but the latter do not. The claim is also made that both nature and our knowledge of it are *stratified* and *differentiated.* That is, having investigated a structure at one level, e.g. chemical reactions, we can investigate the mechanisms underlying and causing this behaviour at a deeper level, e.g. chemical valency, and so on.

Emergent properties, which usually go along with ontological stratification, are not mentioned here, but are defended later in the book (p. 113). At this point Bhaskar is arguing against reductionism in science. He draws the distinction between the physical laws that may underlie the possible behaviours of, say, a machine, and the actual causal factors that lead to it being used in a particular way on a particular occasion. The latter cannot be explained purely in terms of the former, but come from higher-level human or economic systems. He says:

> It follows from this that the operations of the higher level cannot be accounted for solely by the laws governing the lower-order level in which we might say the higher-order level is 'rooted' and from which we might say it was 'emergent'.... In short, emergence is an irreducible feature of our world.
>
> (Bhaskar 1978, p. 113)

He further defends emergence – particularly opposing the cases of society being reduced to the actions of individuals, or mind being reduced to neurophysiology – in *The Possibility of Naturalism* (Bhaskar 1979, p. 97), where he characterizes his position as 'synchronic, emergent powers materialism'. This will be taken up again in the discussion of holistic causality in Section 3.4. Elder-Vass offers an extended discussion of emergence in terms of CR's account of causation (Elder-Vass 2005) and Archer's account of social structure (Elder-Vass 2007a).

The other distinction introduced in the Introduction is between the intransitive and transitive domains of science. The former is the domain of objects of knowledge, such as structures and mechanisms, which are independent of humans, while the latter is the domain of the human production of scientific knowledge.

These concepts can be translated, *prima facie*, almost directly into the language of systems thinking (see Table 3.1): systems forming wholes; a hierarchy of systems with emergent properties; structure and process; and systemic structure and interaction generating observed behaviour. However, when we look more closely we can identify a range of potential differences and distinctions that are worth discussion.

The first point is that Bhaskar is actually quite vague about terms such as 'structure', 'mechanism', 'thing', 'powers' and 'tendencies'. He does not really define them or explain what they might consist of, nor does he make it clear if they are actually synonyms or if there are differences between them. Bhaskar recognizes this in the Postscript to *The Possibility of Naturalism* (Bhaskar 1979), where he responds to some critics. He says that he sees them as a network or family of terms (shades of Wittgenstein) that are interdependent but does not wish to define them more precisely so as to allow readers different ways in to the material.

He does, however, accept that he uses the term 'structure' to mean different things and tries to distinguish between structure and generative mechanism. 'It now seems to me to be better to use the term "generative mechanism" to refer only to the causal powers of ways of acting of structured things' (Bhaskar 1979). This seems to suggest that things have structures and, in virtue of that, possess causal powers, which would be quite usual from a systems perspective. However, the examples he gives do not accord with this. He suggests that a mechanism (his example being the market) may sustain several different structures, and that the same structure (his examples being nation-states or the family) may be reproduced by several mechanisms. First, it is not clear to me the difference between the market and a nation-state such that one is classified as a mechanism and the other as a structure. It also suggests a difference in level – mechanisms underlie and generate structures. But this becomes difficult when we consider that there are in fact many levels – does a structure at one level then become a mechanism for the structures of the next level up?

I would like to suggest that much of this confusion could be avoided if systems terminology were adopted. 'System' would then be the general term for entities, of any type (e.g. physical, social, cognitive, etc.), that populate the intransitive domain. Systems consist of components and their relations which together are characterized as their structure. By virtue of that structure, systems have emergent properties or causal powers or tendencies to behave in certain ways. Systems are stratified: that is, they form nested hierarchies. There are causal relations between systems at a particular level that generate events in the world; and there are causal relations between levels, in that properties or causal

Table 3.1 Comparison of systems concepts with Bhaskar's main earlier and later works

A Realist Theory of Science (1978) and The Possibility of Naturalism (1978)	*Dialectic: The Pulse of Freedom (1993) and Plato Etc. (1994)*	*Systems thinking*
structures, mechanisms, 'things'	totality	systems
	parts/wholes	parts/wholes
powers, liabilities, tendencies	holistic causality	emergent properties
	internal relations	relationships
open and closed systems	open systems	open and closed systems
stratified ontology	recursive embeddings	hierarchy or nesting of systems
emergent properties	emergent properties	emergent properties
intransitive and transitive domains		the observed and the observer
mechanisms generate events		structure generates behaviour or process
	tensed, rhythmic spatial processes	process, dynamics
	absence, negativity, real non-being	
	autopoiesis	autopoiesis
	transformative agency	soft systems, second-order cybernetics
		positive and negative feedback relations
		boundaries

powers of systems at one level, combined through their enduring relations, generate the emergent properties of the systems at the next level up.

One could still use the term 'generative mechanism', which does have valuable connotations, within the context of CR's retroductive methodology. Beginning with some particular events or observations that require explanation, we propose a particular (possibly unique) combination of systems, interacting together in certain ways, that would, if it existed, generate the observed events. We could give this ensemble the term generative mechanism. This is quite similar to Senge's archetypes (Senge 1990), mentioned above, which are particular sets of feedback relations that give rise to certain, common patterns of behaviour. An example of such an archetype is known as 'success to the successful'. Two systems, e.g. universities, compete for a limited resource, e.g. good students. If university A gains a better reputation, for whatever reason, that starts an upward reinforcing feedback loop for A leading to more and more success, and a downward loop for B leading to less and less success, even though both may have been similar to start with. These sort of processes can help explain the relatively wide dispersion, in league tables, of the 'new universities' in the UK that all started from scratch in the 1960s.

3.3.2 Positive and negative feedback

Discussion of feedback, or circular causal relations, leads me to point out that this concept is almost entirely missing from Bhaskar's work. It is not mentioned in any indexes, nor is it an entry in the voluminous *Dictionary of Critical Realism* (Hartwig 2007). There is a brief mention of homeostasis (Bhaskar 1986: 146). Yet, I would argue, it is fundamental in understanding the dynamic behaviour of real-world systems. This omission is partly explicable in that the early books, which we are currently considering, were more concerned with establishing the ontological and structural reality of mechanisms, or systems, rather than with analysing their actual behaviour. There is more consideration given to processes in the later, dialectical, works which we will discuss in Section 3.4.

3.3.3 Processes and events

The next issue to discuss is the concept of 'event', which is central to Bhaskar's model. One of the primary distinctions is that between the enduring causal mechanisms and the temporal events that they generate, leading to the distinction between the domain of the Real and the domain of the Actual. Yet the whole notion of an event is barely discussed at all, even though it is the subject of significant debate within philosophy – see for example the *Stanford Encyclopaedia of Philosophy* entry (Casati and Varzi 2002). For our purpose, I would suggest that an 'event' has two essential characteristics – that it is located at a particular point or interval in time and space rather than being an on-going process or relationship; and that it involves some kind of change to a system or situation, for if nothing changes there is no event. In fact, there is an exception to this last point in the case of an absence. As we will see, for CR absences can be causes, and so the absence of something that *was expected* (e.g. the payment of a bill) could be seen as an event which then has causal effects.

Considering first the time element, the implication is that there must be a start, an end, and some duration; the point is that these are entirely relative to the systems under consideration. Although we tend to think of events in relation to our human time frames, e.g. births, deaths and marriages, in principle they are not absolute but entirely relative. Cosmic events, such as the death of a star, may take millions of years, whereas quantum events occur in nanoseconds. Even on a human scale, events may take a few seconds, a few months or even a few years. The point I want to make is that events are not given to us as things in themselves; rather they must be carved out of the on-going flux of activities and occurrences according to some criterion or interest. And what turns up as events, as opposed to enduring tendencies, depends very much on the time scale that is adopted. If we observe the economy, the credit crunch may be seen as a single event if we take a ten-year perspective, but it may be seen as an enduring tendency generating events of its own in a weekly perspective.

We must second consider the content of the event. We have said this must be a change, or else there would be no event, but a change in or of what? Surely there is nothing that can change other than the entities and structures (i.e. systems) that constitute the Real in the first place. There cannot be events as somehow ontologically distinct kinds of things. Thus events are nothing other than the changes that occur to and within entities and structures. These may be changes to an entity – i.e. it could gain or lose powers, or even disintegrate – or they may be interactions between entities that lead to certain outcomes or outputs. What is crucial, again, is the time frame over which observations occur. The shorter the time frame, the more aspects of the situation that will be fixed or unchanging (structure in systems terms, enduring mechanisms in Bhaskar's); the longer the time frame the more that will become variable and changing (process in systems terms, events in Bhaskar's). So, in other words, the distinction between enduring mechanism and event, or structure and process in systems terms, is ultimately relative to the time frame adopted.

Bhaskar does get close to this conceptualization in a brief section within *Scientific Realism and Human Emancipation* (Bhaskar 1986, p. 215) where he says:

> their concerns fuse ... in the study of process where structure meets events; that is in the study of the mode of becoming, bestaying and begoing of a structure or thing.... Process is not an ontological category apart from structure and event.

From a systems perspective, perhaps, ontologically there are only systems – process is the change to a system which can be sliced up into a series of events.

From a systems perspective this all points to the role of the observer. When conducting some sort of analysis or research, decisions have to be made about the level of the analysis (e.g. organization, department or individual worker), the boundaries of the analysis (narrow or broad), and the time frame. The particular decisions made by the observer (i.e. the analyst or researcher) will determine what shows up as events to be explained rather than as explanatory generative mechanisms. While Bhaskar recognizes the general role of human activity in the production of knowledge through the transitive dimension of science, I would suggest that CR does not pay sufficient attention to the role of the actual scientist or researcher in a specific piece of research. It is the researcher(s) who, based on their own particular interests and predispositions, carve out the object of scientific enquiry by defining both the time frames and the boundaries of the investigation (the domain of the Empirical).

3.3.4 Boundaries

The concept of 'boundary' is itself a central one within systems thinking but is not dealt with by Bhaskar. Arguably, the concept of a 'system' existing within an 'environment' is the foundation for systems theory, and yet what is it that separates a system from its environment? – the system boundary. In fact, defining a system in terms of its components and their relations is effectively to delineate its boundary. Or, put the other way, in order to define a system it is necessary to define its boundary. Thus the drawing of a boundary is in fact the most primitive systemic act that one can perform (Spencer-Brown 1972). However, as soon as we move away from very simple physical objects whose boundary is uncontentious (and that is often the metaphor that Bhaskar has in mind) the decision as to what constitutes the boundary, and thereby what is defined as the system, becomes complex and observer-dependent.

Even with essentially physical systems, there can be many different ways of conceptualizing a system – consider for example a central heating system, or the human body. In these examples we can see that systemic thinking involves more than the simple recognition of particular objects. It begins with a particular phenomenon to be explained or purpose to be achieved. It then requires a degree of conceptualization, rather than mere perception, to characterize an appropriate system in terms of components, relations and boundary. The boundary may in part have a material embodiment, but generally it will simply represent a distinction or demarcation between that which has been selected as part of the system and that which is not. This does not mean that the boundary is purely arbitrary, or is wholly a construction of the observer. It rests on the components and relations that exist independently in the intransitive domain, even though it is selected by the observer. This is demonstrated by the fact that the observer may *get it wrong*. Knowledge is always fallible, and the real world will soon let us know if our choices of components, relations and boundaries do not in fact yield the appropriate behaviour. To quote a well-known management cybernetician, Stafford Beer,

> A system is not something given in nature, but something defined by intelligence.... We select, from an infinite number of relations between things, a

> set which, because of coherence and pattern and purpose, permits an interpretation of what otherwise might be a meaningless cavalcade of arbitrary events. It follows that the detection of systems in the world outside ourselves is a subjective matter. Two people will not necessarily agree on the existence, or nature, or boundaries of any systems so detected.
>
> (Beer 1966, pp. 242–243)

A detailed discussion of the difficulties of defining boundaries in different domains can be found in Chapter 7. The point for CR is that this is a very real issue in actual research projects.

3.4 Systemic concepts in dialectical critical realism

With the move to dialectical critical realism (DCR) there is a tremendous proliferation of terms and arguments. The main underlying structure of DCR is known as MELD and has four aspects: first Moment, second Edge, third Level and fourth Dimension, as shown in Table 3.2.

1M, the first Moment, is characterized in terms of non-identity, that is that things are not all the same, but involve many degrees of differentiation and stratification. 1M includes most of the distinctions from the early philosophy – transitive and intransitive, real/actual/empirical, emergent powers, stratification, generative causality, and mechanisms and events. In systems terms, this is all about structure.

2E, the second Edge, is characterized in terms of negativity and absence, that is, that the world consists as much of things that are not present as things that are; or rather that things that are present, or do occur, only do so against a background of things that are not. This aspect brings in change and development, for it is the need to fill an absence, or equivalently to absent an unwanted constraint or problem, that brings about occurrences and events. In systems terms, this is all about process.

3L, the third Level, brings in many more systemic constructs. It is, in fact, characterized as totality, that is holism and holistic causality. It brings in parts/wholes, inter-relations and inter-activity, recursive embeddings and reflexivity. In systems terms, this is about holism and emergence.

4D, the fourth Dimension, is concerned with human agency, that is, emphasizing that people are causative agents who can bring about change in a purposeful way.

3.4.1 Holistic causality

Let us first consider holistic causality (3L). Bhaskar makes clear what he means by this. It occurs when a *complex coheres* (in our terms a system behaves) in such a way that:

1 the totality, i.e. the form or structure of the combination, causally determines the elements, and

Table 3.2 Overview of the MELD categories (a more comprehensive table is presented in the *Dictionary of Critical Realism*, Hartwig 2007, p. 296)

	1M: first Moment	*2E: second Edge*	*3L: third Level*	*4D: fourth Dimension*
Formal principle	• non-identity (structure)	• negativity (process)	• totality (holism)	• praxis (agency)
Motifs	• critique of anthropism/epistemic fallacy • transitive/intransitive • real/actual/empirical	• absence and ills • axiology of freedom • contradiction and constraint	• holistic causality • reflexivity	• transformative praxis
World characterized by	• intransitivity • stratification • transfactuality • emergence • control and change	• real non-being • negation • process and transition: tensed, rhythmic, spatial processes • priority of negative over positive	• emergence • wholes • internal relations • inter-activity	• human agency • intentionality • autonomy
Critiques of	• actualism: generally reductions to the here and now • anthropism: seeing being in terms primarily of human being	• ontological monovalence: a purely positive account of reality	• ontological extensionalism: a denial of internal relations or causal necessity	• disembodiment • de-agentification • reification of social structure
Dialectics	• superstructuration, stratification and emergence • generative powers, mechanisms and events • structure and agency	• critique • transformative practice	• totalization • centre-periphery • part-whole • recursive embedding	• praxis • hermeneutic struggle

2 the forms or structures of the elements causally co-determine each other and so causally co-determine the whole (Bhaskar 1994, p. 77).

If the word 'determines' sounds overly deterministic, he accepts that the term should include other, weaker, relations such as conditioning, limiting, selecting, sustaining or enabling (Bhaskar 1993, p. 126). In considering this formulation, Part II is quite unexceptional. It is the basic systemic notion of emergent properties in which the behaviour and characteristics of the whole are generated by the structure of the parts and their relationships. However, Part I is much more controversial, even within systems thinking itself. For this suggests that in some way the whole, as a whole, influences or affects its parts, what is often called 'downward causation'.

There are several potential problems here. The first is what we might call a logical concern over levels of organizations, or relations between parts and wholes (mereology). We have seen that systems form nested hierarchies in which the parts and their relations at one level give rise to the properties or behaviours of the whole at the next level up. This local-to-global causation is generally accepted except among strong reductionists. However, in what sense can a whole be said to interact with its own parts? Surely it only interacts with other systems at its own level? Cars as wholes interact with roads and other cars, not with their own engines.

The second problem is the philosophical one of microphysical reduction, i.e. that ultimately physics is seen as a closed and complete system of physical events (Kim 1998). With upward causation it is possible to argue that if the lower level generates the higher level, states at the lower level correspond to states at the higher level (supervenience) which could at least in principle be explained in terms of them (Meyering 2000). However, downward causation would violate this principle and mean that there were genuinely necessary causal elements at levels beyond the physical.

A third potential problem is that holistic causality can easily be interpreted as a version of functionalism. Indeed, classical functionalism, if it existed, can be seen as a clear case of holistic causality – the parts of a system such as a social system actually come into being because the functions they perform are necessary for the maintenance of the whole. This issue is discussed by Bhaskar (1986, p. 142).

These concerns need to be dealt with. One way is through systems theory, in particular complexity theory (or non-linear dynamical systems), which has evolved a language extremely close to Bhaskar. Much of the debate about downward causation is at the level of the mind and its relations to the brain. Thompson and Varela (2001), for example, characterize emergence as follows:

> A network, N, of interrelated components exhibits an emergent process, E, with emergent properties, P, if and only if:
>
> - E is a global process that instantiates P and arises from the nonlinear dynamics, D, of the local interactions of N's components.

- E and P have a global-to-local ('downward') determinative influence on the dynamics D of the components of N.
- E and P are not exhaustively determined by the intrinsic properties of the components of N, that is they exhibit 'relational holism'.

(Thompson and Varela 2001, p. 420)

The form of causation envisaged is one in which states of the whole system (called 'global order parameters') affect the possible states or behaviours of the components by constraining or affording particular paths or patterns of activity. So there is a 'reciprocal causality' in play in which the components interact directly and locally, generating and sustaining the behaviour of the whole, while the whole sets the control parameters and boundary conditions for the components. Thompson and Varela (2001) give general examples such as autopoiesis (Maturana and Varela 1980) and the immune system (Varela *et al.* 1988) as well as specific neurophysiological examples, such as epilepsy, where it is possible to show conscious thought affecting nervous activity. Equally, we can use the example of social systems within critical realism. Here, social structure (or system) is only instantiated through the activities of social agents, but at the same time the social structure of roles and practices conditions the activities that agents can undertake (Archer 2000a; Mingers 2004a).

This approach has the potential to deal with all three of the problems discussed above. If it can be demonstrated empirically in some domain such as the mind or the social world that downward causation does in fact occur, then that refutes the first two objections. The issue of functionalist explanation is too complex to deal with here (see the entry in *Dictionary of Critical Realism*), but certainly Maturana and Varela have always maintained that their theory of autopoiesis is non-functionalist in that circular or reciprocal causality either happens, as a matter of fact, or it does not, in which case the system disintegrates. They also accept the possibility of conflictual processes within a system.

3.4.2 Absence and negativity

One of the major developments along the route from the early work to DCR was the incorporation of absence and negativity as one of the underlying presuppositions. Against the prevailing worldview that deals only with what positively occurs or exists, Bhaskar maintains that it is the absent or the negative which has priority, for it is only against this that the positive stands out or happens. Bhaskar highlights four categories of absence:

1. Simple or ontological absence, i.e. that some thing or event that is expected does not occur or does not exist. Such absences can have causal effect and therefore 'exist' in the same way as other things. He calls them 'de-onts'. The instrument that is not to hand, the bill that is unpaid, or the appointment that is missed all have causal effects.

2 Absence as a verb, that is, absenting something or negating something, e.g. draining water or removing dirt; or absenting an absence, e.g. removing a need or want by fulfilling it.

Developing from these are

3 'process-in-product' whereby a process (e.g. shopping) leads to an absence (e.g. money in the bank).
4 'product-in-process' whereby an entity or structure (e.g. poverty, lack of money) exercises its powers in producing an absence (necessities of life).

This is interesting from a systems thinking point of view because it is *not* something that is generally discussed or considered in the modern literature and yet is clearly of great importance. In fact, its significance was recognized by some: it can be seen as the basis of cybernetic explanation, as Bateson, one of the founders of cybernetics, observed.

> Causal explanation is usually positive. In contrast to this, cybernetic explanation is always negative. We consider what alternative possibilities could conceivably have occurred and then ask why were many of the alternatives not followed, so that the particular event was one of those few which could, in fact, occur.
>
> (Bateson 1973b, p. 375)

A similar idea is at the heart of Luhmann's (1990b) theory of social communication in which a message acts as a trigger or selector from among the many responses or replies that could be generated – it selects that which is presenced from among all the other absent possibilities. We can also see the importance of absence in the idea of control by feedback. The feedback system (e.g. a thermostat) is always trying to close a gap (absent an absence) between the desired state of the system and the actual state of the system (Wilden 1977).

So, here could be a valuable contribution from CR to systems in terms of bringing to the forefront that which is generally absent, namely the concept of absence itself.

3.4.3 Autopoiesis

Autopoiesis means, literally, systems that are *self-producing* or *self-constructing*. In traditional systems theory, systems were seen as open, transforming inputs into outputs. Biologists Maturana and Varela (1980; 1987) developed the concept of autopoiesis to explain the special nature of living as opposed to non-living systems. Autopoietic systems are closed and self-referential – they do not primarily transform inputs into outputs, instead they transform *themselves into themselves*. The components of an autopoietic system enter into processes of production or construction to produce more of the same as necessary for the continuation of the system. The output of the system, that which it produces, is its own internal components, and the inputs it uses are again its own components. Autopoietic systems are said to be organizationally closed but interactively open (Mingers 1995b). The paradigm example is a single-celled organism such as an amoeba.

There have been several attempts to apply autopoiesis at levels above biology, in particular to suggest that social systems may be characterized as autopoietic (Luhmann 1982b; Luhmann 1986; Robb 1989b), but this is controversial and remains an open question (Mingers 1992a; Mingers 2002). Elder-Vass (2007b) has specifically contrasted Luhmann's approach with Bhaskar's and Archer's emergentism.

Bhaskar himself uses the term autopoiesis in several ways, although never referring to the original literature. In *Scientific Realism and Human Emancipation* he characterizes the whole process of knowledge production in the transitive domain as 'quasi-autopoietic':

> [Cognitive resources] comprise the transitive objects of knowledge; their transformation is the transitive process of knowledge production; and its product, knowledge, in turn supplies resources for further rounds of inquiry. This imparts to the cognitive process a quasi-autopoietic character.
>
> (Bhaskar 1986, p. 54)

This is actually an interesting potential application of autopoiesis which has been little explored.

In DCR, Bhaskar generalizes this idea to describe the (re)production of the social system as a whole through human activity:

> The activity-dependence of social structures entails its auto-poietic [*sic*] character, viz. that it is itself a social product, that is to say, that in our substantive motivated productions, we not only produce, but we also reproduce or transform the very conditions of our productions.
>
> (Bhaskar 1993, p. 156)

Again, this is a very appealing metaphor, but can we go beyond metaphor and claim that social systems *are*, ontologically, autopoietic (Mingers 2002; Mingers 2004a)? We will return to this question in Chapter 6.

Finally, Bhaskar characterizes emergence itself as being autopoietic: 'In emergence, generally, new beings (entities, structures, totalities, concepts) are generated out of pre-existing material from which they could have been neither induced nor deduced.... This is matter as creative, autopoietic.' (Bhaskar 1993, p. 49.) This is certainly a stimulating idea for debate within both CR and systems.

3.5 Conclusions

Bhaskar, in one of his discussions about absence, uses the example of books in a library and the many forms of absence that give them meaning. This chapter, too, has been about absence – in this case the huge absence in Bhaskar's work of any reference to the domain or literature of systems thinking despite its informing so many of his ideas. This is not said simply as a criticism of Bhaskar but rather as a recognition and invitation for further development.

It may be hoped that this chapter has served three purposes:

- to point out and justify the claim that many of the fundamental ideas of CR have already been developed within the disciplines of systems thinking and cybernetics, and thereby to open up this literature for followers of CR;
- to try and demonstrate that systems thinking has much to offer CR in terms of providing clearer articulations of its concepts, and also other concepts such as circular causality through positive and negative feedback loops, that could be useful for CR;
- to suggest that CR can also be beneficial for systems thinking partly by providing a more rigorous philosophical underpinning that systems thinking lacks, and also by its development of particular concepts such as absence/negativity.

At the least, I hope that this chapter may open up dialogue and debate between the two disciplines.

Notes

1 I shall use these terms interchangeably. The meaning of other related terms such as cybernetics, general systems theory or holism will be brought out in the text.
2 Good sources for overviews of the history of systems are Capra (1997); Checkland (1981); Hayles (1999); Heims (1993). There is an interesting and very detailed timeline at the *American Society for Cybernetics* (2006).
3 Although many of the ideas were foreshadowed by Bogdanov (1980, orig. 1922) in his work on 'tektology', which was not widely known at the time.
4 These ideas formed the basis for Prigogene's work in the 1970s (Prigogene and Stengers 1984) on dissipative structures, for which he gained the Nobel prize.
5 The term *kybernetike* was used by Plato and Aristotle to mean the art of steering or governorship.
6 Which is distinct from 'second-order cybernetics', to be discussed later.

Part II

Ontological issues

4 Explanatory mechanisms

4.1 Introduction

Chapter 3 demonstrated many ways in which critical realism (CR) drew on systemic concepts. In this chapter, I will focus particularly on the idea of explanatory or generative mechanisms, a concept that is at the heart of CR and is also currently of great interest in the philosophy of science.

For many decades, the concept of explanation within the philosophy of science was presumed to revolve around the idea of universal laws. Events were to be 'explained' in terms of being instances deduced from general covering laws, which were themselves developed through some form of induction from observable empirical examples. This was formalized in what became known as the deductive-nomological (D-N) model developed by Hempel (1965). However, in more recent years the limitations of this approach have become ever more obvious and an alternative has been generating much interest and support. This approach eschews universal laws in favour of particular 'mechanisms' that causally generate the phenomena of interest to the scientist (Gerring 2007; Glennan 1996; Glennan 2002; Machamer 2004; Machamer *et al.* 2000; Salmon 1998a; Symons 2008). Apart from avoiding many of the problems besetting the D-N model, especially concerning induction, the idea of mechanisms fits much better with the actual practices of scientists (Bechtel and Abrahamsen 2005) and, as we shall see, with explanations in everyday life.

As might be expected with a major new development, there are many issues to be debated, and indeed arguments about how the term mechanism should be conceptualized. In this chapter, we will contribute to the understanding of mechanistic explanations by bringing in theoretical ideas from two domains that are highly relevant to this approach, and yet that have so far been little discussed. These domains are those of systems thinking and cybernetics, and the philosophy of critical realism. The former is an obvious choice. Many of the papers on mechanistic explanation explicitly couch their models in terms of complex systems in which a variety of component parts interact with each other to form a mechanism that then has particular behavioural or emergent properties. However, these papers tend to make no reference to the huge literature on systems thinking and cybernetics that developed, originally in biology, ecology, and information science, from the 1920s onwards. This provides a strong body of conceptual and theoretical work on which the mechanistic viewpoint can be built.

Less well known, perhaps, is the philosophical approach of critical realism that was described in Chapters 2 and 3, which, as we saw is a comprehensive and sophisticated post-positivist paradigm that has at its heart the idea of generative causality via causal structures or mechanisms which possess powers or tendencies to behave in particular ways. The actual and empirical events that occur in the world are then seen as resulting from the interactions and interplay of these structures and mechanisms. We will show that many of the issues and problems of the mechanistic view of causation have already been encountered, and to some extent addressed, within critical realism.

4.2 Deduction, induction and abduction: the logic of mechanistic explanation in critical realism

Our starting point in explaining critical realism's view of mechanisms is actually the work of the pragmatist philosopher C.S. Peirce (1931–1958). One of Peirce's many contributions was the development of the logic of 'abduction' (or 'retroduction' in CR terms) as opposed to deduction or induction (Psillos 2009).

Consider the following syllogism (an example from Peirce 1931–1958, 2.623[1]):

General law:	All beans in this bag are white.
Particular case:	All these beans come from this bag.
Conclusion:	All these beans are white.

This syllogism, which can be classified as AAA-1 (known by the mnemonic 'Barbara'), is valid and is an example of deductive inference. Given a general law or rule we can deduce a particular consequence from it.

The three elements can be re-arranged as follows:

Context:	All these beans come from this bag.
Empirical observations:	All these beans are white.
General law:	All beans in this bag are white.

This syllogism can be said to capture the logic of induction – from particular instances we induce a general conclusion. In terms of pure logic it is invalid, since there could still be beans in the bag that are not white but were not selected, but it obviously has utility as a practical mode of inference.

These elements can also be re-arranged as follows:

(Unexpected) observation:	All these beans are white.
Possible cause:	All beans in this bag are white.
Explanatory hypothesis:	All these beans come from this bag.

This syllogism is also not valid in a logical sense – some other reason could explain why all the beans are white – but it has quite a different character from the previous two. Peirce called it 'abduction' or 'retroduction'. In his Fifth Lecture on Pragmatism, Peirce said, 'Abduction consists in studying facts and devising a theory to explain them' (1931–1958, 5.145) and in the Sixth Lecture, he said 'abduction is the process of forming an explanatory hypothesis' (5.171). So, with the process of abduction we begin with some particular occurrence or event, usually one that is unexpected or does not conform to current theories; and we then take an imaginative leap to think of some theory or explanation which might account for the event. This is neither an induction from the examples nor a deduction from the rule, but rather an explanatory or exploratory hypothesis as to why the situation might have occurred.

Abduction is essentially the same as what is currently known as 'inference to the best explanation' (Lipton 2004), a phrase originated by Harman (1965). This argues that, given two theories, if one explains more or better than the other then that provides warranted grounds for choosing it. It has been criticized by Laudan (1984) and van Fraassen (1989).

Abduction is the point where novelty, innovation and creativity enter the scientific method, as indeed they must. With deduction, we get nothing more than the consequences of the premises – but where did they come from? With induction, we just get a generalization from the observations we have made – but how do we know they are all that matters? However, with abduction we get explanation and the possibility of new knowledge.

Peirce (1931–1958, 2.781) recognized that actually all three modes were necessary for successful science. We begin with an unusual or puzzling phenomenon, C, and try to hypothesize something, A, which would account for the existence of the phenomenon; this is abduction. Then, second, we explore the consequences of A. If A is in fact the case, what other consequences would follow that we might be able to observe or test? This is deduction. Finally, we try and observe whether these consequences do in fact happen, which would thus confirm our explanation. This is a form of induction.

This mode of reasoning is also at the heart of critical realism, which adopts an approach to causality that is known as 'generative causality' (Mingers 2000a). In distinction to positivism or empiricism, which adopts the impoverished Humean version of causality as nothing more than a constant conjunction of events, CR holds that there is a stratified external reality in which the occurrences and events we experience (the 'Actual') are the result of, or generated by, the interaction of underlying structures and mechanisms (the 'Real'). These mechanisms may be physical, social or conceptual, and they may be observable or unobservable. CR's methodology (DREIC), which involves retroduction, is very much along the lines of Peirce's abduction (Hartwig 2007, p. 195):

D An unusual occurrence or anomaly is observed and Described.

R Retroduction is applied putative causal mechanism(s) are hypothesized which, *if they existed*, would account for the occurrence or anomaly.

E Hypotheses are Eliminated where possible.
I The correct mechanism is Identified.
C Changes are made to existing theories.

The main difference from Peirce is that the account is couched in terms of mechanisms and structures that have causal powers or tendencies to bring about changes in the world. Bhaskar does not explicitly define what he means by 'mechanisms', and sometimes he refers to them as 'structures', but they can be characterized by the following:

- Mechanisms exist in a real, ontological sense independently of how they may be known or described by observers. They are stratified, in the sense of depth or hierarchy, and they may be physical, social or conceptual. They may be observable or unobservable. Their existence is judged by a causal rather than a perceptual criterion – i.e. that they have causal effects in the world.
- Mechanisms are relatively enduring in respect of the events that they cause, but their absolute time scale may vary immensely. They have powers or tendencies, by virtue of their structural properties, to behave in particular ways or have certain effects. These powers may not be exercised all the time (perhaps needing to be triggered), or they may be exercised but have no effect because of the countervailing actions of some other mechanism. Through their interactions, mechanisms generate the actual occurrences and events of the world, only some of which are observed or noted empirically (Bhaskar 1979, p. 170). Thus a mechanism may be said to consist of a structure of inter-related parts together with the powers or tendencies that the structure possesses.
- Social structures or mechanisms have different properties or characteristics from physical ones (Bhaskar 1979). First, they only become manifest at all through the activities that they govern. That is, social structures cannot be directly observed; they exist only virtually, as a set of practices or roles which govern or enable social activities – think of language as an example. Through these activities the structures become reproduced or indeed changed and transformed. Second, social structures rely to some degree on the knowledge and understanding of social actors, who must be aware that they are doing a particular activity and know how to do it. Third, social structures are localized in time and space in the sense that they belong to particular cultures at particular times rather than being universal, apart perhaps from extremely general ones such as the human ability to use tools or language. Finally, social systems are inevitably open (rather than being able to be closed as in a laboratory experiment) and hence, in principle, unpredictable.

We should also point out, and this will be discussed more below, that critical realism stresses the importance of the negative/absent as well as the positive/present as causally efficacious (Bhaskar 1993).

4.3 Parts, wholes and boundaries: the basis of systemic thinking

In Chapters 1 and 3 we traced, in some detail, the origin and development of systems thinking and it concepts. These too can be seen as fundamental to the essence of mechanistic thinking. Mechanisms are, indeed, no more and no less than systems of interconnected components that display emergent properties, and which form nested hierarchies. Such systems and components must have some forms of boundary, and this will be explored in more detail in Chapter 7.

4.4 Issues in mechanistic explanation

Illari and Williamson (2011) provide a useful overview of some of the issues involved in the philosophy of mechanistic explanation. We will use their overview to structure this section, in which we try to show that systems thinking and critical realism can shed some light on these problems. Mario Bunge (2004a; 2004b) has explicitly adopted the idea of mechanisms within a systems thinking approach.

4.4.1 The nature of mechanistic explanation

Mechanistic explanation is clearly in contrast to the covering law model for several good reasons, and this has been one of the main arguments of CR against various forms of positivism and empiricism (Bhaskar 1978; Groff 2011). Positivism, in the form of the D-N model of explanation and resting on a Humean view of causality, involves a double reduction. It first reduces the domain of the real – enduring entities and structures that have causal powers – to the domain of the actual – particular events that actually occur (ignoring absences, i.e. events that might have occurred but for some reason did not). And then, it reduces the domain of the actual to that of the empirical, i.e. those events that happen to be observed and can be measured. From this impoverished base, it does no more than re-describe the data in the form of a mathematical law, with no greater concept of causality than constant conjunctions of events (Craver 2006).

In contrast, CR only begins with empirical observations; it then goes beneath the surface to try and explain what underlying mechanisms could, if they existed, produce the events that in fact occurred, or did not occur. CR has a stratified ontology: the Real, which consists of enduring mechanisms and structures, including human beings and social systems; the Actual, which are the events caused or precluded by the interacting mechanisms, and which themselves of course can have causal effects; and the Empirical, which is the subset of the actual experienced and observed by human beings. Bhaskar supports this view both on logical (transcendental) grounds that we will discuss later, and on more pragmatic ones that echo those philosophers supporting a mechanistic approach. In particular, these more pragmatic grounds reflect that this approach properly provides an *explanation* for events rather than simply a redescription for them

(Glennan 2002); that it accords with the actual practices of scientists (Bechtel and Abrahamsen 2005); and, again to be discussed later, that general or universal laws do not exist in many domains, especially the social sciences (Cartwright 1983). Chakravartty (2005) defends this view against several long-standing objections.

Illari and Williamson suggest that a second aspect of mechanistic explanation is a distinction in the literature between epistemic and ontic types of explanations. Salmon (1998a) originally drew this distinction, suggesting that an epistemic form of explanation, such as that of Hempel (1965), was essentially an inferential argument to the effect that the events to be explained were to be expected given general laws and the initial conditions. On the other hand, an ontic explanation (sometimes called a physical explanation) is one which shows how the events have resulted from causal patterns and regularities. These may be of two types – constitutive, where the events result from the properties of a specific mechanism, and etiological, where they are the outcome of a chain of events. More generally, an epistemic explanation is motivated by a desire to improve human understanding, and is therefore constrained by the knowledge and understanding of the audience. An ontic explanation concerns the actual causal mechanism and its effects, whether or not it is properly understood. A similar distinction has been made between actual mechanisms in the world, and scientists' descriptions and models of those mechanisms (Bechtel and Abrahamsen 2005; Glennan 2002; Machamer *et al.* 2000). As Illari and Williamson say, 'These differences exist because in the epistemic sense of explanation it is the *description* of the mechanism that explains, while in the physical sense, the *mechanism itself* does the explaining' (2011, p. 823, my emphasis). This distinction can be seen as part of a wider differentiation made by Bhaskar concerning what he calls the *transitive* (epistemic) and *intransitive* (ontic) dimensions of science (Bhaskar 1979, Chapter 1). It has long been argued, especially from within the sociology of science, that science is a human activity or practice much like any other, and therefore that the results of science reflect such human practices as much as they do the object world. This argument can be taken to have significant sceptical conclusions, for example in the 'strong' sociology of knowledge programme (Bloor 1976) or various forms of postmodernism. Bhaskar accepts that indeed much of science is a human activity or production, which he calls the transitive dimension, but maintains strongly that there is also an intransitive dimension to science which consists of the objects of knowledge that are independent of us, or at least of how we describe or know them.

Thus, the transitive dimension involves all the human activity of producing knowledge or, perhaps better, transforming previously existing knowledge, and is therefore inevitably local, temporal and partial. We have to accept that knowledge can never be perfect, or even be 'proved' to be correct. It is always fallible or epistemically relative, but this does not mean that all theories are equal, or that there are not rational grounds for choosing between them. One reason for this is precisely the externality or ontological independence of the objects of knowledge in the intransitive domain. Such objects do not have to be physical, but can be social, cognitive or even linguistic. An academic paper is produced in the transitive dimension, but once published it becomes an intransitive object of knowledge able to be discussed or referenced.

Bhaskar also has a multivalent model of truth which is relevant in this context (Bhaskar 1993; Mingers 2008). This involves four levels or degrees of truthfulness. The lowest level (*normative-fiduciary*) is when one simply accepts the truth of what someone says on the grounds that they are a reliable or knowledgeable person who should 'know' what they are talking about (e.g. a scientist or expert). Clearly this is very common in everyday life. The second level (*adequating*) is truth that is based on sound evidence or justification of some kind rather than mere belief. Both these levels are in the transitive domain, and thus relate to the epistemic approach above. The third level (*referential-expressive*) is like a weak correspondence theory relating theories or models in the transitive domain to their intransitive objects.

The final level (*alethic*) is somewhat controversial (Groff 2000) for it moves the truthmaker entirely into the intransitive domain. There is no longer a correspondence between different domains, for it is the truth of things in themselves, and their generative causes, rather than the truth of propositions. It is no longer tied to language, i.e. it is no longer necessarily linguistic, although it may be expressed in language. It seems to be very akin to the ontic view of explanation that Illari and McKay described, as discussed above. One could perhaps say that the experience of toothache generates its own alethic truth – we do not need to describe it or compare it with something else, we merely experience it to know its truth.

4.4.2 The reality of mechanisms

It may seem obvious, but if mechanisms are to be the core of scientific explanation, then it is necessary that mechanisms be 'real', that is, at least some must have an independent existence and be responsible for the phenomena that they explain. Clearly there are extreme sceptical arguments that would question whether we can take anything to exist, including ourselves – the age-old argument addressed by Descartes and Husserl. We will not consider these, but there are nevertheless important issues that need to be addressed, particularly the debate over 'theoretical entities', the possibility of non-physical mechanisms such as social and cognitive systems, and the question of functionalist explanation.

First, theoretical entities – committed positivists and empiricists have denied the legitimacy of unobservable, theoretical entities within scientific theories on the grounds that they were not observable or measurable, and so could not be assumed to exist. This is to adopt a perceptibility criterion for existence. Such a criterion is clearly against the practices of working scientists, who routinely hypothesize the existence of unobservable mechanisms and then set about trying to observe them or their traces (witness the billions spent on the large hadron collider and the search for the Higgs boson, believed now to have been discovered), and it is clearly against the possibility of mechanisms in the non-physical domain. Bhaskar (Archer *et al.* 1998; Bhaskar 1978; Bhaskar 1989) has proposed several arguments against this view, and the related Humean view of causation as constant conjunctions of events. The main form of argument, which is also employed in a slightly different way by Cartwright (1999), is a transcendental argument à la Kant.

Transcendental arguments (Stern 2000) take the form:

Premise 1: There is something, X, that we experience or agree about.
Premise 2: X could not be experienced if Y were not the case.
Conclusion: Y must be the case.

For Bhaskar, the X that we experience and agree on is experimental scientific activity (within natural science), both its successes and failures. In conducting a laboratory experiment we, scientists, bring about (or fail to bring about) a particular effect under certain controlled conditions. Effectively, we engineer constant conjunctions of events which do not, in fact, happen regularly at all. Then, however, we find that these effects can also be brought about outside of the lab, in open rather than closed conditions. For this to happen it must be the case that (Y) causal laws are more than simply constant conjunctions – there must in fact be enduring structures or mechanisms that are distinct from the events they generate, and occur both inside and outside the laboratory. 'On this view, laws are not empirical statements but statements about the forms of activity characteristic of the things of the world. And their necessity is that of a natural connection, not that of a human rule.' (Bhaskar, in Archer 1998, p. 34.) Clarke (2010) analyses the transcendental arguments of both Bhaskar and Cartwright, concluding that while neither succeeds entirely, neither should be rejected out of hand. Bhaskar himself developed some of his ideas from previous work by Harré and Madden (1975) on the notion of causal powers,[2] and it is also possible to relate these arguments to the Aristotelian approach to causality (Pratten 2009). Criticisms of the transcendental arguments were discussed in Section 2.4.

The second issue concerns the reality of non-physical mechanisms such as social, psychological or informational systems. In fact, most of the philosophical literature on mechanisms restricts itself very much to the natural sciences such as biology and chemistry; the ontology of social structures has always been highly contentious in social science (Mingers 2004a). We cannot cover such a debate here, but we will highlight three related issues: i) whether we can accept non-physical and non-observable mechanisms (or systems, or structures) as having ontological reality; ii) whether there are social structures or mechanisms over and above the actions of individual people; and iii) whether the dependence of social mechanisms on peoples' understanding of them somehow compromises their reality. For critical realism, the answer is clearly 'yes' to the first two and 'no' to the third (Bhaskar 1979, 1989, 1997).

CR utilizes a causal criterion for existence and a transcendental argument, as discussed above. Entities do not have to be physical or directly observable to be real; they only have to be causally efficacious. This means that concepts, ideas, rules and social practices, for instance, are no less real for being unobservable. We can also apply the transcendental argument to social structures. Our experiences of the social world cannot be explained purely in terms of individual actions – it must be the case that there are social mechanisms in operation that exist before, and over and above, particular individuals, and that necessarily enable our social activities. Some obvious examples are language, the banking system and the use of money more generally, or the legal system. Considering language, it is something that pre-dates us, that we have to learn in childhood, and yet it then enables us to communicate with people we have never met. It is, in a general sense, a human construction for it would not exist without us, and yet none of us individually can change or develop it. For these experiences to occur it must be the case that language exists as an unobservable structure or mechanism separate from its embodiment in individuals' nervous systems and from its instantiations in actual language use.

For Bhaskar, society exists as an object in its own right, emergent from, but separate from, people and their activities, and with its own properties. Society always exists before individuals, who do not therefore create it but only transform or (re)produce it. Nevertheless, society is necessary for social activity and it only *exists* in virtue of that activity. Society therefore conditions social activity and is either maintained or changed as an outcome of that activity. Equally, human action (praxis) is both a conscious production, i.e. intentional bringing about of purposes, and an unconscious (re)production of society. Society is an 'ensemble' of structures, practices and conventions, where structures are relatively enduring generative mechanisms that govern social activities. While emphasizing the ontological reality of social structures, Bhaskar recognizes that they have significantly different properties from physical mechanisms as was mentioned above. In particular: (i) social mechanisms do not *exist* independently of the activities they govern; (ii) social mechanisms cannot be *empirically identified* except through activities; (iii) social mechanisms are not independent of actors' *conceptions* of their activity; (iv) social mechanisms are *localized* to particular times and cultures. Despite these differences they are still suitable subjects for scientific theorizing, even if theorizing about them leads to particular epistemological difficulties. The explicit use of 'mechanism' as an explanatory device is growing in social science; see, for example, in history Steinmetz (1998), in organizational research Anderson *et al.* (2006), and in politics Gerring (2007).

A third issue raised by Illari and Williamson (2011) is that many of the writers on mechanisms (e.g. Bechtel and Abrahamsen 2005; Craver 2007; Darden 2006; Glennan 2002) presume that they must fulfil some *function* within a wider system, so that a mechanistic explanation must be a functionalist explanation. This is problematic partly because of the long-standing debate about the validity of functionalist explanation, especially in the social sciences (Salmon 1998b), but also because the specification of a function would seem to depend on a description of the wider system which may in turn depend on the perspective of the observer.

Our view is that insisting that a mechanism must fulfil some function in order to be a mechanism is neither necessary nor legitimate outside of the domain of humanly designed systems. Certainly modern systems theorists do not accept it. For example, the biologist and neurophysiologist Humberto Maturana (1970b; 1975b), who developed the concept of autopoiesis (self-producing systems) to explain the fundamental nature of living entities, was clear that his use of mechanistic explanation was non-teleological; so was Antony Giddens (1984), with his concept of structuration as a theory of the reproduction of social systems. Mechanisms arise historically through some particular chain of events, perhaps involving chance. They have effects, and it may well be that these effects, often in combination with the effects of other mechanisms, may give rise to behaviour that is self-sustaining or that contributes to a wider system (another mechanism at a higher level). In this sense, they may be seen, *after the event and by an observer*, to play some functional role. However, the actual operation of the mechanism still occurs in terms of its own structure and local interactions:

> The organization of a machine ... only states relations between components and rules for their interactions and transformations, in a manner that specifies the conditions of emergence of the different states of the machine which, then, arise as a necessary outcome whenever such conditions occur. Thus the notions of purpose and function have no explanatory value in the phenomenological domain which they pretend to illuminate.
>
> (Maturana and Varela 1980, p. 86)

Not only is functionalist explanation unnecessary, but it may be incorrect, since mechanisms may have effects which are dysfunctional or non-functional as far as some wider system is concerned. Obvious examples are cancerous cells or insurgency, both of which are still the result of organized mechanisms. So it is right and proper to characterize mechanisms in terms of their components, relationships and emergent powers and properties, and then observe their behaviour in interaction with other kinds of mechanisms. The issue about a mechanism playing different roles in different systems will be discussed in the section on localization.

4.4.3 Must mechanisms be local?

If we move, now, to the actual nature of mechanistic explanations, or more precisely the nature of the mechanisms that are postulated in such explanations, we find another potential problem in the literature – that such mechanisms are generally said to be 'local' in a spatio-temporal sense (Bechtel 2001; Craver 2007; Hall 2004). Illari and Williamson (2011) identify three possible difficulties – the functional individuation of mechanisms, the existence of non-physical mechanisms, and omissions or absences – but still conclude that 'it is a genuine feature of mechanisms that they are local' (p. 833). The argument of this section, from a system perspective, is that while many mechanisms (especially physical ones) are indeed local, in a physical space sense, this is not a necessary characteristic of a mechanism. Rather, a mechanism or system operates within a state-space dependent on the characteristics of its constituting components, which may or may not be physical.

There seems to be a general agreement among those advocating mechanistic explanation (e.g. Bechtel and Abrahamsen 2005; Gillett 2007; Glennan 1996; Wimsatt 1994; Woodward 2002) that the postulated mechanisms are what we have called above 'systems', i.e. groups of component parts (or mechanisms) that interact together to create a particular effect or phenomenon which is to be explained. They generally form hierarchies, with emergent properties at each new level. The implication of this conception – particularly in the case of physical phenomena, which is what most of these authors discuss – is that the mechanism and its effects are localized in a physical sense. The size of the locality is hugely variable, from quantum to astronomical scale.

From a systems viewpoint, we would translate the idea of localization into one of boundary. In order to identify a system (or mechanism, using the two equivalently) we need to distinguish what elements constitute the system as opposed to its environment, and this means specifying the system's boundary. This is not a straightforward task because of the variety of different types of system, as can be seen from the following points (discussed further in Chapter 7).

A boundary is that which separates or demarcates that which is part of a system from that which is not. It may be physical or non-physical, actual or conceptual. In the case of physical systems, we can distinguish:

- Edges and surfaces that are the limit or extent of a system, e.g. a pool of water.
- Enclosures, where there are specific boundary components that keep in that which is included, and/or keep out that which is excluded, e.g. a football, a fence, or a membrane such as the cell wall.
- Demarcations, where the system is physical but not necessarily contiguous in space, e.g. the solar system or a central heating system.

In general, systems can be conceptualized in different ways, generating different boundaries; and the components of a system may be parts of multiple systems. For example, a central heating system could be conceptualized as a flow-of-water system (pipes, radiators, valves, water supply), a flow-of-energy system (gas, boiler, water, air), or a flow-of-information system (difference between actual and desired temperature, difference in thermostat setting, difference in degree of heating). These examples show that systemic thinking involves more than the simple recognition of individual objects, as we saw in Section 3.3.4.

So far we have considered primarily physical systems where the idea of spatial localization may be seen to be necessary, although even there non-physical elements such as information are often involved. However, as we move away towards conceptual or social systems we need to characterize them in terms of the space of interactions of the system itself, which may well not be physical space.

Let us consider as an example the nervous system as a system (or mechanism) in its own right, separate from, although obviously part of, a body (Maturana 1970b, 1980a; Maturana *et al.* 1995; Varela *et al.* 1991). An organism without a nervous system, such as an amoeba, acts in response to chemical changes in its immediate local environment. Its outer wall is essentially both its sensory surface and its effector surface. However, in the nervous system cells have become specialized in two particular ways. First, they have developed lengthy extensions called dendrites that connect them to many other, sometimes distant, neurons. This leads to a separation of sensor from effector and the possibility of a transmission of difference or disturbance. Second, they have developed a generalized response medium – electrical activity – into which all forms of sensory/effector interactions can be translated. A third development is that of internal neurons that connect only to other neurons and that form the vast majority of developed brains. These effectively intervene in the direct connection between sensor and effector and are the basis for cognition, language and ultimately self-consciousness.

The effect of this is that the development of the nervous system opens up a whole new domain of interactions beyond the purely local physico-chemical ones of amoebae. The nervous system allows the organism to interact with the *relations* or *differences* between events, rather than simply with the events themselves. Neurons develop that are triggered only by particular combinations of other neurons, which represent particular configurations within the environment. Thus, although the nervous system is a physical system and does have physical interactions, its domain of interactions *as a nervous system* is states of relative neuronal activity triggered by relations and differences in its environment and acting back on that environment. Such interactions cannot be localized to the brain. For example, in a mobile phone conversation with someone physically distant, differences in sound are transmitted through the phone system to differences in sound at the receiver, differences in brain activity, differences in sound, etc. The whole forms a system in which spatial location is not a necessary or defining feature. This view is related to theories such as radical enactivism (Menary 2006) and the extended mind (Clark and Chalmers 1998).

The nature and boundaries of other non-physical systems, especially social systems, is controversial and will be discussed in more detail in Chapter 6 (Archer 1995; Bhaskar 1979; Giddens 1984; Luhmann 1995; Mingers 2002, 2004a), but we will give one illustration. It is very common in the commercial world to talk about the 'market mechanism' as a particular type of economic process, and we would argue that it can, indeed, be seen as a social or economic mechanism. It is a particular form of (largely unregulated or uncontrolled) trading where buyers and sellers interact fairly directly, and prices change in response to the balance between supply and demand. Historically, this mechanism actually did operate locally in a physical market, located in space and time, and lasting for a certain duration. However, today we can see that trading, especially in financial or commodity markets, is highly attenuated, being conducted electronically throughout the world and 24 hours per day. It does not make sense to talk about its locality, but it is still important to delineate the boundaries of such a system – i.e. what constitutes the system and what constitutes its environment, even if the answer may depend in part on the observer and their purpose.

4.4.4 Absences and omissions as causes

Finally, in this section we shall discuss the question of omissions or more generally absences as causal elements of mechanisms. Illari and Williamson (2011) suggest that absences may be a problem for mechanistic explanation on the grounds that they may not be local, a condition that the authors consider necessary, as discussed in the previous section. Torres (2009) proposes a revision to the mechanistic models of Glennan (1996) and Machamer *et al.* (2000) to more properly account for negative or absent causes.

We have already argued above that we do not consider localization as a necessary condition for a mechanism, and so we do not consider the fact that absences sometimes cannot be localized to be a problem. Rather, we would argue strongly that varieties of absence often lie at the heart of causal mechanisms; both critical realism and systems thinking, particularly cybernetics, support this view, as discussed in Section 3.4.2.

4.5 Conclusions

The purpose of this chapter has been to demonstrate that many of the issues and debates within the recent philosophy of causal mechanisms have already been much discussed within the literatures of critical realism and systems thinking, and, moreover, that the conclusions reached there may well be useful in the philosophical discourse about mechanisms.

After a very brief review of the literatures of critical realism and systems thinking, a range of issues concerning mechanisms were discussed. In particular, it was argued that:

- We need to distinguish between the events that occur (and do not occur) and that are to be explained (the Actual) and the underlying, enduring structures and mechanisms (the Real) that, through the operation of their powers in interaction, causally generate these events. We should also distinguish between the transitive domain of science (which is epistemic), in which theories and knowledge are humanly produced, and the intransitive domain of the independent objects of our knowledge (which is ontic).
- We can accept that the reality of mechanisms (or systems more generally) may be non-physical and/or non-observable. The ontological criterion should not be perceptibility but causal efficaciousness. Thus, social mechanisms (e.g. 'society'), informational mechanisms (driven by information), or

cognitive mechanisms (e.g. ideas or motives) all have causal effects and may thus be part of explanatory theories.

- Mechanistic explanation does not have to be a form of functionalist explanation.
- Mechanisms do not have to be localized in a purely physical sense, although they need to be bounded, or able to be demarcated, within their space of interactions.
- Absences and omissions may be causes and thus may legitimately be part of mechanistic explanations.

This chapter has only been able to skim the surface of the many possible connections between mechanistic explanation and systems thinking and critical realism, but it is hoped that this discussion will demonstrate the value of such an engagement for both sides.

Notes

1 References to Peirce's *Collected Papers* (1931–1958) are in the form (vol.para).
2 Although Harré has distanced himself from critical realism, especially in respect of social structures (Harré 2002).

5 Emergence, generative powers and causality

5.1 Introduction

This chapter will focus in more detail on some of the basic concepts that underlie both systems thinking and critical realism. Critical realism and, more generally, mechanistic explanation are fundamentally opposed to the Humean concept of causality as constant conjunctions of events. In its place is a conception of generative causality based on the emergent properties and powers of interacting systems.

As we saw in Section 4.4.1, positivism and empiricism are underpinned by a Humean account of causality. From a CR point of view, this involves a double reduction. First it reduces the domain of underlying generative mechanisms or systems to the domain of actually occurring (or not occurring, although positivism tends to ignore this) events; then it reduces the domain of all events to include only those that happen to have been observed or measured (the empirical). It then essentially redescribes the observed data in terms of mathematical formulae or models (Craver 2006). In contrast, CR has a stratified model of explanation. Events occur as the result of interactions between mechanisms or systems, and the mechanisms have particular properties or powers which are an emergent feature dependent on their structure or composition.

The first section of this chapter will discuss emergent properties and causality in general – what exactly are emergent properties and how do they underlie causality? The second will look more specifically at what we mean by properties and whether they are different from powers or tendencies. If they are, what is the relationship between them? The next section will consider specifically the relations between the *actual* events that occur and the *real* domain of generative mechanisms – what exactly do we mean by 'events'? Finally, we will consider the problem of abstracting or generalizing putative laws or regularities from particular patterns of events.

5.2 Emergence and causality

Let us begin with emergence. This is a long-standing concept at the heart of the reductionism/holism debate, and also a central core of systems thinking. Elder-Vass (2005, 2007a, 2010) provides a very good explication and defence of the notion of emergence, and emergent properties, within the context of CR; in the main, we will follow his analysis. We will differ on certain aspects, as we will see later.

Emergence has been an important issue within philosophy for many years, for example in the philosophy of science, where it is central to the reductionism/holism debate, and in the philosophy of mind concerning the relations between mental and physical phenomena.[1] The term was first used by Lewes (1875), developing work of Mill (1843). Broadly, 'emergence' represents the idea that new properties or entities emerge from the combination of lower-level ones in such a way that they are not reducible to them. A corollary of emergence is that the world must be stratified into a hierarchy of types of systems, each one developing out of the lower-level ones. This in turn leads to the development of distinctively different sciences – chemistry, biology, psychology and sociology – which cannot be reduced one to another (Wimsatt 1994).

Emergence is also the key idea of systems thinking, as often expressed in the phrase 'the whole is greater than the sum of the parts', for it is precisely this that allows us to maintain that systems exist over and above their component parts. This was a key part of the development of systems thinking within biology as part of the fight against vitalism as an explanation of living systems (Broad 1925; Ritter 1919).

In more recent times, emergence has taken on a greater prominence in several domains. As we have seen, Bhaskar has made it central to critical realism, which he described, in the beginning, as 'stratified, emergent powers materialism' (Bhaskar 1978). The rise of complexity theory or science (Byrne 1998; Cilliers 2000; Kaufmann 1995; Mainzer 1997) has strengthened anti-reductionist arguments. In many complex or chaotic domains it is clear that their dynamic behaviour is highly unpredictable given even good knowledge of the properties of their components. Small changes in initial values can lead to widely divergent outcomes, and hard-to-predict dynamic effects such as tipping points make it impossible to explain behaviour by reducing it to the properties of lower-level components (Bedau 1997).

Philosophers such as Kim (2006), O'Connor (O'Connor and Wong 2005) and McLaughlin (1997) have argued for a strong ontological 'supervenience'. Property B is said to supervene on property(ies) A if the occurrence of the A properties (or their combination) always generates the B property. In other words, systems that are alike in their basal properties will also be alike in their emergent properties. For example, whenever hydrogen and oxygen combine in appropriate quantities the result will be a substance with the properties of water. Equally, if we have a substance identical in all it properties to water then it must be composed of hydrogen and oxygen. We shall discuss this further below.

5.2.1 Emergent properties

This section largely follows Elder-Vass (2005, 2007a, 2010). In general terms an emergent property or power (using the terms synonymously for the moment) is a property that is possessed by an entity or a whole that results from, but is not possessed by, any of its parts. In many discussions of emergence in the literature there is a tendency to assume that the entities are physical rather than, say, social or cognitive, although that is not at all necessary and Elder-Vass does include a business organization as an example. Originally, Compte (1986) and Durkheim (1951, orig. 1897) used the idea of emergence to establish sociology as an autonomous discipline above psychology. To avoid inadvertent reification, I will try and use the term 'system' to cover all possible types of entities.

We then need to distinguish between synchronic and diachronic emergence. Synchronic emergence describes the relations between systems and their components that exist at a point in time, or that remain unchanging over a period of time. Diachronic emergence refers to the way systems develop over time, e.g. a flower emerging from a bud. The emergence discussed in this section is generally synchronic emergence. A similar distinction will be drawn later, in the section on causality, but there we need to involve both perspectives in explaining actual changes and occurrences.

In terms of parts and wholes, we follow the general systems approach already outlined in earlier chapters. At any particular level, a system is composed of a variety of parts or subsystems. These in turn can be analysed into their own subsystems and so on to the most fundamental forces. Subsystems enter into a variety of relations with each other and, taken together, they and their relations form the structure of the system. The system *is* its structure, taken as a whole, but we have to recognize that the domain of its interactions as a whole with other systems is distinct from the domain of the interactions of its parts (Maturana and Varela 1980).[2]

Elder-Vass distinguishes between systems with *emergent* properties, which have 'significant' structures, and systems that have only *resultant* properties because they have no significant structure – called 'heaps' (Laszlo 1972). Examples of heaps might be relatively arbitrary constructs, such as 'all the grains of rice in China', which do not have any actual relations; or properties of systems which are resultant rather than emergent since they are simply the sum of the properties of the parts – for example the weight of some grains of rice (which is a property of both the parts and the whole) or the average height of a group of people (which is a property only of the whole). Note that we cannot dispense with the entities or systems in favour of simply free-floating properties. The properties must be properties *of* some particular system.

This latter point leads Elder-Vass to adopt a compositional view of emergence. By this is meant that a system has its emergent properties by virtue of its structure – that is, its component parts and their relationships. It does not require any other components, *that do not necessarily belong to it*, to exhibit its properties. This essentially leads to a view of discrete, bounded systems, each with a set of properties or powers dependent only on its own necessary set of parts. This is very much the view that was discussed in Chapter 4 on mechanistic explanation. I think that although it is true in many situations, there are two problems with it. First, is the whole question of boundaries. While it is reasonably easy to identify or agree on the boundaries for physically discrete objects, there are many types of systems, especially non-physical ones such as social systems, where there are no clear boundaries, or where different boundaries may be drawn by different observers or for different purposes. We will discuss this whole issue in much greater detail in Chapter 7.

The second concern, which is recognized by Elder-Vass, is that some powers of a system can only be enacted in combination with some other appropriate system or environmental interaction. Elder-Vass draws a distinction between the existence of a property and its actual realization. Bhaskar is also explicit on this point – he claims that powers may exist but not be exercised, needing something to trigger them; or that powers may be exercised but not actually have any effects because they are counteracted by some other, external causal power. Let us consider some examples of this.

If you cut yourself, this triggers many reactions in the body such as closing down the blood vessels, clotting the blood and then clearing out damaged cells. This is an example of powers existing but not generally being exercised until triggered. A second example: when you get ill you often get a temperature – the body produces excess heat. If you take medication, this can reduce the temperature to normal; when the medicine wears off, the temperature returns. This is an example of powers being exercised (the body is still ill) but being counteracted so that they are not apparent. Another example is given by Elder-Vass which seems to reflect a different situation – that is the power of water to put out a fire. This, he says, is a power that exists but that is not realized unless there is a fire to put out. This seems to me to raise a problem. It is certainly true that water has certain properties, being a cold, inflammable liquid, which means that if there is a sufficient quantity it could put out a fire. But does that mean we would want to say that putting out fires is actually a property of water? Would we not rather say that it has certain properties such that, in a particular situation, there will be certain effects?

What I am trying to suggest is that we should distinguish between properties that are intrinsic to the system, regardless of its interactions,[3] and properties (or perhaps powers) that only come about in particular sets of circumstances or in interactions with other systems.[4] Consider the example of colour. Many people would say that a property of a ripe tomato is that it is red. However, this is actually very contentious[5] and the prevailing view is that colour is *not* a property of the object but a subjective property of the perceiver. All the surface of the tomato does is to absorb or reflect certain wavelengths of light. It is only when this is received by a colour-responsive nervous system that the experience of colour occurs. So at best it is only a property of the interaction between perceiver and perceived, but should perhaps be in the category of a property of the perceiver that is triggered by the tomato.

We will consider this issue again when we discuss whether there are in fact differences between properties and powers.

To return to the description of emergence and causation, we have now established that a system has emergent properties because of its significant structural relationships. This structural composition is the 'mechanism', to use Bhaskar's term, that generates or explains its particular properties. To explain a higher-level system in terms of its lower-level components and relations is not reductive in an eliminative sense. We are not trying to remove the higher-level system, but we are trying to explain it. Such an explanation could therefore be said to be explanatory reductionism. The structure explains the emergence in the sense that whenever that particular structure occurs, so will those particular properties. However, this is not true in reverse. There may be other structures that will generate the same properties – for instance, heat may be generated by many different forms of interaction within different types of system.

5.2.2 Causality

We have already to some extent discussed (in Sections 2.2.1 and 4.4.1) the empiricist, Humean view of causation as purely a constant conjunction of events and we are now ready to develop more fully a non-Humean version of causality.

The fundamental difference is that Hume conceives causation as being no more than repeatable sequences of events. We can never go beneath or behind the events to understand why they occur, one after the other: 'experience only teaches us how one event constantly follows another; without instructing us in the secret connexion, which binds them together, and renders them inseparable' (Hume 1967, orig. 1750). In contrast, realists such as Bhaskar believe, 'contra Hume, that causal relations are relations of natural or metaphysical necessity, rather than of contingent sequences' (Groff 2008, pp. 2–3). This is the central point of Bhaskar's distinction between the Actual and the Real – the Actual is the domain of events that occur (and absences that do not – as discussed in Section 3.4.2), and the Real is the domain of mechanisms and systems which, through their interactions, generate the events. In terms of causation, this means that we need a distinction between 'real' causation and 'actual' causation. Real causation concerns the powers and tendencies that complex systems have to affect the world *whether or not these powers and tendencies are actually realized on any particular occasion, or indeed ever*. These are the properties or powers that we have just analysed in terms of their emergence from the structure of the mechanism or system. They are real in that they do or could have causal effects on the world if and when they are actualized, and whether or not they are actually observed by anyone. This relates to Bhaskar's criterion for existence – causal efficacy rather than perceptibility. Thus a plane has the power to fly whether or not it ever does, and it has that power by virtue of the properties of its components (e.g. they are the right shape and strength) and their relationship (e.g. the wings are connected appropriately to the fuselage).

In contrast, actual causation concerns the particular events that happen and the way they result, causally, from the interaction of the various powers of the systems involved. Actual causation is nearly always multiple in the sense that it is the often unpredictable outcome of a range of different interactions both synchronically and diachronically. This is where science, and particularly the laboratory, differs from everyday life. In the lab, it is possible to control many of the factors involved in order to home in on a particular interaction in a particular context. In these circumstances, constant conjunctions of events can be produced artificially – X is always followed by Y because there is nothing to interfere with it or countervail it. Bhaskar sometimes distinguishes between these two forms of causation in talking of 'retrodiction' when trying to determine antecedent causal events, and 'retroduction' when trying to determine what underlying mechanisms may generate them (Bhaskar 1994, p. 30).

We should also mention what may be seen as the obverse of causal powers – that is, liabilities. In many cases, the exercise of a power implies the existence of a corresponding liability in another system. Thus the power of a knife to cut through cloth implies that cloth has that liability with respect to the knife. If the cloth were in fact armour, then it would not have that liability. So a liability is just another form of causal power although in some sense a negative one – the power to be affected in a particular way by some other causal power.

Thus, events in the real world happen, or do not happen, through the interaction of a variety of different systems (causal mechanisms), each with their particular causal powers and liabilities. However, as we have seen, systems are composed of components; while systems have a domain of interactions as a whole, they also interact with each other, at the same time, through their components. So, when I meet someone I respond to them as myself and as themselves (as wholes) but I also shake their hand and smile with my face, and perhaps sub-consciously notice something about their body language. All of these complex interactions, at different hierarchical levels, come together to generate the actual event that occurs.

One obvious question is: are the interactions between the parts any different in principle from those between the wholes? The answer is 'no', since the components are themselves wholes albeit at a different level. To the extent that there is any difference, it is merely one of persistence. The interaction between me and the other person is relatively fleeting in comparison with the interactions between all of my components as a person, since my components are bound together over a much longer period of time because of the morphostatic connections which bind together me as a system.

There are two other issues that we need to deal with to round out our account of causality, although as they have already been discussed earlier in the book we will only summarize the conclusions here. These are the notion of downward causality, and that of absences as causes.

So far we have been discussing what might be called upward causation; that is, the way in which the parts of a system, with their properties and relationships, causally generate the emergent properties of the whole. This is upward in a hierarchical sense. But in many systems it also appears that there is downward causation, i.e. the whole has causal influence on the parts. In Section 3.4.1 three problems with this concept were identified. The first is a logical problem of levels – can a higher-level system or whole interact with its lower-level parts? Surely it can only interact with other systems at its own level. Second, strong reductionists see the concept of downward causation as violating the principle that higher-level states should be explainable only in terms of lower-level states, ultimately in terms of physics. Third, the notion of downward causation is a form of functionalism which is generally seen as unacceptable, especially with regard to social systems.

The approach to avoiding these problems, which can be seen to work with regard to both the mind/brain and the individual/society relationships, is to argue that states of the whole system (global order parameters) affect or condition the possible states and paths of behaviours of the component parts without entirely determining them. So, for example, social structures such as roles or practices both enable and constrain the activities of social actors even though these structures only exist by virtue of such individual activity.

Absences as causes have already been discussed in Section 3.4.2. For traditional positivism or empiricism it is only things that actually exist, and indeed that can be perceived and measured, that could be causes of other things. However, as Bhaskar strongly argues, in fact there are many absent entities or events that can have causal consequences – the appointment that is missed, the bill that is unpaid, the food that cannot be afforded all leave their marks on the world.

In this section (5.2) I have outlined a general account of emergence and causation that accords both with critical realism and with the philosophy of mechanisms discussed in Chapter 4. In the main I have followed Elder-Vass (2010). However, there are particular areas where I either disagree with Elder-Vass, or wish to develop the account in a different direction. The first is powers and properties – Elder-Vass sees these as synonymous whereas I do not, as I will discuss in Section 5.3. Next I will clarify more precisely the difference between the Real and the Actual, and in particular the nature of an event (Section 5.4) and the distinction between synchronic and diachronic causation (Section 5.4.1). Finally, in this chapter, I move away from explaining particular events or instances of causation to the problem of abstracting from particular instances more general causal regularities – what might be seen as a modern form of induction – through the work of Pearl and Woodward (Section 5.5).

One other limitation of Elder-Vass's account is that, although his book actually deals significantly with social systems, I believe that he has a tendency to reify them – that is, to treat them as if they were like material things. In particular, he tends to tacitly assume that they are generally localized with well-defined boundaries. This is a result of his view that systems are ultimately compositional. It also relates to the issue of the localization of mechanisms discussed in Section 4.4.3. From my perspective, there is a major problem in determining the boundaries of any system that is more than a fairly simple, self-bounded, physical objet and especially of any non-physical system such as a social system. This is a very complex problem and will be treated to a chapter in its own right (Chapter 7).

5.3 Powers, properties and tendencies

Within the critical realist literature, and particularly Bhaskar's own work, terms such as 'power', 'property' and 'tendency' are often used somewhat interchangeably, and there is also a debate within philosophy as to the difference, if any, between powers and properties. Within the positivist tradition, particularly following Hume, causation was seen in terms of universal laws applied to rather inert objects. However, as we have been exploring, within the newer mechanistic approach causality is a result of the intrinsic powers or properties of active, natural systems (Chakravartty 2008). The question addressed in this section is: are there significant distinctions to be made between the various terms used, or should we just take them as synonyms?

Let us first consider 'property' and 'power'. We have seen that properties, as in emergent properties, are characteristics, attributes or behaviours of systems that are causally generated by the structure of the system, which consists of parts and their relations. Thus, 'incompressibility' is a property of water. The term 'power' is generally taken to concern the causal effects that a system can have in the world, thus, 'water has the power to put out a fire'. Several authors argue that powers are essentially the same as properties

For instance, Elder-Vass (2010, p. 17) says, 'Properties and powers may therefore be regarded as synonyms'. Chakravarrty (2008, p. 154) says, 'I will use these terms [causal power, disposition, capacity] synonymously, to refer to properties of things in virtue of which they behave in particular ways in particular circumstances'. And Bhaskar, although not generally very explicit about it, says 'The ontological bases of powers are just the properties that account for them' (Archer *et al.* 1998, p. 72).

However, not all agree. Fleetwood (2009) suggests that there are those who give primacy to powers, e.g. Mumford (2008), Chakravartty (2008) and possibly Bhaskar; those who give primacy to properties, e.g. Bird (2008) and Mackie (1977); and those, including Fleetwood, who see neither as primary but both as intrinsically linked, e.g. Cartwright (1997) and Shoemaker (1997). I follow the latter group in that I argue that properties and powers are distinct, as we will see below, but that both come into being at the same time as characteristics of the particular thing or entity.[6]

A property is a characteristic or behaviour that is causally generated by the structure of the entity. It exists as a property separately and independently of any other entities or systems. It may or may not be observed; it may or may not be exercised at any particular time. In contrast, a power is the ability to generate or cause a particular outcome *to or with some other entity or system*, within a particular context (or perhaps 'environment' in systems terms). The essential difference is that a power describes the possible effects of a property on other systems.[7] What we then find is that the relation between the two terms may be multiple. One property may generate one power; several properties may be needed to generate a power; or one property may generate several powers. Let us illustrate with the example of a knife.

Something is a knife (or can be used as a knife) if it has two essential properties – it has a sharp edge and it has a certain degree of hardness. Both are necessary for it to function normally as a knife. For example, a knife made of butter would not cut very much unless of course it was frozen to give it a greater degree of hardness. This is where context enters. The degree of hardness and sharpness necessary depend on what is to be cut – even a steel knife will not cut diamond.

So, in this example we can see that the power (to cut) depends on two properties, and also depends on the particular context. We can also see that an entity need not have been designed to be a knife to have the power of cutting – a sharp flint can also cut, as it has the appropriate properties. The properties themselves are generated by the particular structure of the object as in the manner analysed by Elder-Vass.

We can also have the situation where a single property can be said to generate several powers dependent on the context, in particular what it interacts with. Consider for example the property of being acidic. Put simply, acids are chemicals that tend to lose a hydrogen ion in solution; bases tend to gain an ion. The strength of the acid depends on how easily or fully the chemical loses ions. Thus there is one emergent property, although it can be generated in a variety of chemical structures (i.e. different chemicals), but this can display a variety of powers: for example, acids in the stomach break down food; acid rain damages the environment; vinegar (acetic acid) preserves food; acids burn the skin. In each case essentially the same property, when manifest or exercised in a particular context, i.e. interacting with different systems, produces different effects, i.e. powers.

We can, finally, have a single property linked with a single power: for example, the weight or mass of a hammer (property) means that it can bang in nails (power).

To summarize our view of a power:

1. A power is the ability of a system to produce a particular effect in a particular context or in interaction with some other system.
2. This is dependent on the system having a certain property or properties.
3. These properties are emergent from the structure (components and relations) of the object.
4. This is provided that the context is in fact realized (and that there is no countervailing action).

When expressed like this, these four conditions can also be seen as similar, if not identical, to the four traditional Aristotelian causes, as shown in Table 5.1.[8]

5.4 Events: the Actual and the Real

Within this approach to causation there needs to be a clear distinction between the underlying enduring causal mechanisms and the events that they cause. For Bhaskar, this distinction also generates the domains of the Real and the Actual. We have already discussed, in Section 3.3.3, the fact that the notion of an 'event' is not actually developed in Bhaskar's work, and in that section we tried to clarify the idea.

An event has two essential components – it has a duration, with a start and finish, although this may be long or short; and it must involve a change, or else there is nothing to observe (although this includes the absence of something

Table 5.1 Powers and Aristotle's four causes

Aristotelian cause	*Interpretation*	*Properties*	*Definition of a power*
Formal cause	What makes it a knife?	It is hard and has a sharp edge	The system has a particular property (properties) that generates the effect
Final cause	What is the knife made for?	Cutting	The system is able to produce a particular effect in a particular context or interaction
Material cause	What is the knife made of?	Steel that has hardness and a shape that has a sharp edge	The system has a structure that generates the necessary properties
Efficient cause	What made the knife? OR What made the knife cut X?	The manufacturers OR The person holding it	The particular context or circumstances are actually realized

happening that was expected to). The conclusions we came to were that events were not somehow ontologically distinct kinds of things. Events are nothing other than the changes that occur to and within mechanisms and structures, i.e. systems. These may be changes to a system – e.g. it could gain or lose powers, or even disintegrate – or they may be interactions between systems that lead to certain outcomes or outputs. What is important in defining an event is the time frame over which the observations happen. If the time frame is short (relative to the processes of the system), then most aspects will not have changed and so will be considered structure (in systems terms), or perhaps enduring mechanisms in Bhaskar's terms. The longer the time frame, the more that will have been able to change and so will be considered process in systems terms, or events in Bhaskar's. Imagine observing a lecture theatre. If you observe it daily, then there will be a series of events each day – the various lectures. If you observe it minute by minute, you will see different events – questions being asked, students arriving and leaving. If you observe it only every decade you may see it being built, or being knocked down.

This, I believe, is quite similar to the view of Elder-Vass (Elder-Vass 2005) when he distinguishes between what he calls real causal powers and actual causation. When systems interact, they do so not just as wholes, but also at the various levels of their components. A whole set of interactions occurs at a variety of levels, and these together jointly produce the outcome, i.e. the event. But, Elder-Vass argues, the causality that occurs between the different entities is no different in principle from the causality that occurs between the parts within an entity. The only outward difference is that one is relatively short-lived or fleeting, while the other is relatively enduring or persistent because of the morphostatic or self-maintaining nature of the whole.

So the picture we get is one of ceaseless interaction between different entities at different hierarchical levels. Some of these entities form relatively enduring wholes through their internal morphostatic relationships; these then interact with each other in a more contingent and short-lived way, but the essence of the causal relationship is essentially the same.

> Real and actual causation both therefore appear to be consequences of the same generic type of structural relation: the (diachronic) causal consequences that flow from a given set of entities existing (synchronically) in a given set of relations to each other.
>
> (Elder-Vass 2005, p. 335)

This view is similar, I believe, to that expressed in some of the more recent writings of Bhaskar (2010, 2013) where he is particularly concerned with applied research in open systems. Here he argues that the real world is a complex intermeshing of systems of different types and at different levels, what he calls a 'laminated' system or totality. As such, it will require interdisciplinary research to fully understand its varied aspects.

This leads us to consider synchronic and diachronic causality.

5.4.1 Synchronic and diachronic causality

We can now move to focus in more detail on the actual process of causation. The Humean account talks only of successions of events; the Bhaskarean account, in the main, talks of generative causality from underneath, so to speak. But as the above quotes begins to show, it is in fact necessary to have both of these aspects in a proper account of causality – the diachronic chain of one event influencing the next, and the synchronic relations that generate what actually happens at each event point. These are distinguished by Bhaskar in terms of retroduction and retrodiction. This is a similar distinction to that made by Salmon (1998a), mentioned above, between constitutive and etiological types of ontic explanations.

To illustrate the argument let us imagine a scenario – Bob is in the kitchen and puts down a glass vase on the worktop; Sue is unaware of this and turns around, knocking into the vase; the vase falls to the floor and breaks.

We can begin by looking at this diachronically. We can see the whole as a single event – the vase getting broken – but we can also split it into several separate but related sub-events. Although this split is to some extent arbitrary, nevertheless it is possible because we can see that each sub-event could have had a different outcome, and that if it had the result would have been different. In Table 5.2 we have split it into four sub-events: Bob putting down the vase, Sue turning round, Sue knocking the vase, and the vase falling to the floor and breaking. Each of these sub-events could have had a different outcome, as shown in Table 5.2. Bob could have put the vase down loudly so that Sue knew it was there; he could have put it in another place out of range; Sue could have moved differently and so not come into contact with the vase; it might not have been a vase she came into contact with but the kettle, which would not fall off; Sue might have caught the vase; it might not have broken; and so on.

What we can see here is a series of events each one of which sets up the conditions for the next one but does not wholly determine what the next one will be. The conjunction of all four of the events is a sufficient cause of the vase breaking (assuming no other countervailing events have occurred), i.e. it is an example of material implication; but it is not a necessary cause, as something else could have caused the vase to break, say a cat jumping on the worktop. In terms of the individual events, we can see that one event does not cause, in the sense of determine, the next event. Rather, it sets up the conditions for the next interaction, but what actually happens depends both on the conditions and on the properties and powers of the interacting systems.

Synchronically, we can see (as described in Section 5.3) that at each event particular systems interact, both as wholes and as parts, and generate a particular outcome, which could always have been different. Thus, in the first event the vase was placed at a particular position, and in such a way that it made little noise. Had it been put down in a different place, or more noisily, the later outcomes might not have occurred. When the vase fell off, it broke because of its

Table 5.2 Synchronic and diachronic causality

Diachronic (etiological) causation			
Event 1 Bob puts down vase on worktop	**Event 2** Sue turns around	**Event 3** Sue interacts with vase and it falls	**Event 4** Vase interacts with floor and breaks
Synchronic (constitutive) causation			
The structures of Bob's hand, the vase and the surface are such that the vase can be picked up and put down	Sue turns around not knowing the vase is there	Sue's position and movement is such that she knocks the vase with sufficient force to knock it off	The fragile glass hits the hard floor with sufficient force to break it
Possible alternatives			
Bob might have put down the vase somewhere else	Sue might have heard the vase being put down, or moved somewhere else	Sue might not have actually contacted the vase, or might have done so gently so that it did not fall	The vase might have been caught or made of plastic, or there might have been a chair under the worktop

relative fragility and the force of collision. Had it been made of a different material, or fallen from a lower height, the outcome again might have been different.

In each case, we can see that the result is nothing other than a change in the configuration of the systems involved, whether that is a spatial change or, in the case of the vase, disintegration. The events are changes to the systems or mechanisms, which are, relatively speaking, more enduring.

5.5 Abstracting causal relations – demi-regs

In the example above, we have been looking at the causal explanation of a particular, singular event(s). To some extent the CR approach does tend to presume that this is the situation, with its talk of explanation not prediction and of open systems and the interaction of causal mechanisms, and with its antipathy towards constant conjunctions and universal laws. However, it is clearly the case that the world, both natural and social, is actually highly patterned and repetitive – indeed if it were not, humans and other animals would not exist. So, it is in fact quite natural, and it is certainly very much part of science, to generalize from particular occurrences to regular and repeated instances of similar events. Much of the talk of causation in the scientific literature also concerns the general rather than the particular, for instance, 'what causes cancer?' From a scientific viewpoint, we are not so much interested in why person A got cancer, but in which factors have the tendency to generate cancer, and which may have the tendency to prevent it.

Within CR, the main person to consider this is Lawson (1997, p. 204) in economics. Lawson uses the concept of demi-regularities or 'demi-regs', which are 'a partial event regularity which *prima facie* indicates the occasional, but less than universal, actualisation of a mechanism or tendency, over a definite region of time-space'. This can come about either through a single mechanism that exercises it powers on a regular basis in a particular time-space region, e.g. an active volcano, or through several mechanisms that regularly interact in the same way, e.g. everyone leaving work at the same time and causing traffic jams. These are different from universal laws because they may change over time or space – the volcano may become dormant, or industrial estates and roads may open or close – and they always occur in open dynamic systems which may alter their usual behaviour.

Once we accept the importance of demi-regs, then it is of interest as to how we might spot such regularities and also how we might account for them with causal explanations. If the situation is one in which some form of data can be collected, then this brings in the whole question of the role of statistics within critical realism (Mingers 2006). Although we cannot address this whole topic here (we will look at it in more detail in Chapter 9), I do wish to point to a relatively philosophical approach to statistical causality that I believe is quite compatible with CR in a way that the traditional, Humean form of empiricism is not.

5.5.1 Discovering causal relations

We began this discussion of causality by contrasting the generative mechanism view with that of the Humean constant conjunction of events. Now the Humean view underpins the view of causality that is embedded in modern statistics, as epitomized for example by multiple regression or structural equation modelling. As every statistics course teaches, correlation implies only association, not causality. In other words, from the fact that two variables show a correlation or association of values, one cannot conclude that one variable has a direct causal relation to the other. It may be that both variables are related to some third underlying factor. For example, it has been observed that there is a high correlation between membership of the Church of England and the number of teenage pregnancies in the UK. Of course, one does not cause the other; both are dependent on changes in the underlying population. Or it may be that the observed correlation is simply spurious, a chance occurrence. The same considerations apply to more complex statistical models such as multiple regression. Statisticians, and econometricians such as Hendry (1990), generally accept that statistical causality like this does not translate into the underlying generative causality of critical realism. In their turn, critical realists have tended to be highly critical of statistical modelling and have given the impression that it is antithetical to CR (Bhaskar 1979; Lawson 1997; Manicas 2006; Porpora 1998a; Ron 1999).

However, there have been recent attempts to make the move from traditional statistical causality to generative statistical causality. The basic position adopted for example by Pearl (2000) and Woodward (2003) is quite similar to our account of generative mechanisms,[9] a point already made by Mingers (2006) and Pratschke (2003). Pearl has made this transition personally:

> Ten years ago I was working in the empiricist tradition. In this tradition, probabilistic relationships [e.g. correlation] constitute the foundations of human knowledge, whereas causality simply provides useful ways of abbreviating and organizing. Today my view is quite different. I now take causal relationships to be the fundamental building blocks both of physical reality and of human understanding.
>
> (Pearl 2000, p. xiii)

Pearl's formal position is very clearly stated:

> Nature possesses stable causal mechanisms that, on a detailed level of descriptions, are deterministic functional relationships between variables, some of which are unobservable. These mechanisms are organised in the form of an acyclic structure, which the scientist attempts to identify from the available observations.
>
> (Pearl 2000, p. 43)

The general thrust is good. The idea that there are underlying, stable causal mechanisms and that the scientist tries to identify them from empirical observations fits well with the retroductive methodology, but a considerable amount of unpacking is needed to see just how compatible this approach is with CR.

The starting point is the idea that we have observations of various factors which, we believe, are causally related in potentially complex ways. Pearl also accepts that there may well be other factors involved that we cannot, for whatever reason, observe. The level of measurement may be interval, or it may just be nominal as, for example, on/off or present/absent. On the basis of the observations, we can see that some variables are correlated with each other, and others are independent. The traditional approach of statisticians to correlation, regression and ultimately structural equation modelling (SEM, where one has a set of variables linked by several regression equations) is to assume that the relationships are only statistical and that one cannot infer actual causality from them. Pearl's major philosophical contribution is to argue that one can interpret them causally, and to provide methods for deducing, with a high degree of confidence, actual causal relationships from sets of statistical data.

The first step is to represent the relationships graphically, based on the data and on one's knowledge or beliefs about the underlying causal mechanisms. The graphs consist of nodes (the variables) and links (the relationships) which may be unidirectional arrows, bi-directional arrows, dotted lines representing possible underlying variables that affect two nodes, or non-directional lines. Some of the graph may represent putative unobserved factors and relationships. Such graphs usually, but do not necessarily, involve a time dimension in the sense of one factor preceding another. Where such a graph does involve arrows, it is called directional. Where the arrows all go in the same direction without cycling back (as in a feedback relation), it is called acyclic. For the most part, Pearl is concerned with directional acyclic graphs (DAGs). An example is shown in Figure 5.1.

Each direct relationship in the DAG is assumed to be based on a stable underlying mechanism that is essentially independent of the other ones represented in the graph. Such a graph can be interpreted functionally or as a Bayesian network. In the former case it is assumed that each relationship is

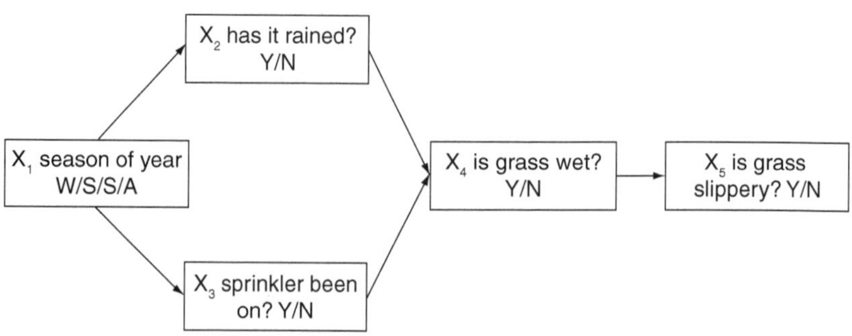

Figure 5.1 Directional acyclic graph (DAG).

a deterministic function, although the particular form of the function may well be unknown. Any uncertainty is due to unobserved variables. In a Bayesian network, each relationship is assumed to be inherently stochastic and new evidence or observations are used to update estimates of the probabilities (Williamson 2005).

Such a network representation provides important and useful properties: a variable depends only on those immediately impacting it, not others earlier in the network (the Markovian property). Thus X_5 (is the grass slippery?) depends only on X_4 (is it wet?); any effects of X_2 and X_3 (the reasons why it might be slippery) are fully accounted for by X_4 (ignoring any other possible causes of slippery grass). The different parts of the network are independent of each other because of the assumption of stable underlying mechanisms. This means that one part can change without affecting other parts, and that interventions can occur without affecting the structure – e.g. the value of a variable can be fixed or adjusted, not just observed, while other relationships remain the same. Thus its effects can be more easily determined. Of course, if it has unexpected effects then this suggests that the structure of the diagram is incorrect.

Pearl's general approach is that we try to develop a DAG that is consistent with whatever data we have, which will primarily be in terms of observed dependencies and conditional independencies[10] between the variables (in probabilistic terms). Given that we can include hidden or unobserved variables (what is then called a latent structure), there are in fact an infinite number of possible models, so Pearl calls on Occam's Razor in terms of the minimality principal: chose the model(s) with the least number of hidden variables that still fits the data.[11] Once this has been done, it is then possible, in some circumstances, to infer direct causation of one variable on another: 'Given P^ [the joint probability distribution], a variable C has a causal influence on variable E if and only if there exists a directed path from C to E in every minimal latent structure consistent with P^' (Pearl 2000, p. 46). In other words, if all possible latent structures consistent with the data do have a direct link between two variables, then such a link must in fact be a necessary causal link. Such inferred causal links between X and Y will always involve at least one other variable (Z) – it is not possible to infer causation knowing only the two variables involved. This third variable can be interpreted as a form of control or intervention, an interpretation that leads to a particular view of causation in general that is espoused by both Pearl and Woodward (2003). This is known as the manipulationist or interventionist view of causation (Dawid 2000; Dawid 2002; Spirtes *et al.* 2001). It says that we only have a proper explanatory account of causation (as opposed to a purely descriptive one) when we have identified factors or variables which, *if they are controlled or changed*, bring about changes in the outcome to be explained. On this view, then Z could represent such an intervention on the causal variable X. If we are merely observing, rather than actively intervening, then the Z can be seen as nature's manipulation of X.

Although most of this section has focussed only on diagrammatic methods of causal inference, this approach in general can be applied to mathematical statistical analyses such as regression and, in particular, to structural equation modelling (SEM) (Pearl 2009).

I will move on now to considering how well this approach fits with CR. The overall position seems to be very compatible with a strong ontological commitment to the existence of stable causal mechanisms and with a methodology that tries to explain empirical data in terms of these mechanisms, whether observable or not. Indeed, at one point, while discussing the analysis of counterfactuals, Pearl does use the term 'abduction' (p. 206). I think the main problem is the very limited form of representation, i.e, maps of quantitative variables. As Bhaskar has made clear, the social world is different from the natural world in many ways, not least the fact that much of it, based on meanings, cognitions and interpretations, is not able to be measured. This does not make it any the less causal. In many ways the social world is more like a novel or play than a machine – we have to interpret and understand what is meant by particular communications and actions, not just count them. Trying to fit this into a quantitative DAG would be much like doing statistical analyses of the frequency of word usage in a Shakespeare play – it tells you something but in many ways misses the whole point.

It is even harder to expect that we will be able to accurately construct functional relationships between all the variables. In practice, SEM tends to routinely assume that all relations are linear or can be simply transformed to linearity. There are in fact other forms of diagramming approaches which could also be used, although without the rigorous assumptions of DAGs it would not be possible to reconstruct causal relations in the same way. One of these methods is influence diagrams in the form of multiple cause diagrams as used in system dynamics (Sterman 2000). These specifically embody circular feedback connections – something banned from standard *acyclic* DAGs. These result in simulation models rather than statistical regression models. The Bayesian interpretation of DAGs is also perhaps more flexible than the functional version proposed by Pearl (Williamson 2005).

A second concern, perhaps more with Woodward than with Pearl, is that the manipulationist/interventionist approach to causality is much more epistemic than ontological. To say that we have causality only when we can intervene to manipulate the values of a particular variable and then observe its effects seems to be much more about how we might *know* whether or not something is causal rather than whether or not it actually *is*. Certainly the majority of Woodward's book concerns our causal *explanations* rather than causes themselves, although he does accept a modest form of realism in saying that 'it requires only that there be facts of the matter, independent of facts about human abilities and psychology' (Woodward 2003, p. 118).

5.6 Conclusions

In this chapter we have developed the notion of generative causality, which underlies the concept of mechanistic explanation. Mechanisms generate events because they have causal powers: that is, they have characteristics or properties that allow them to have effects in the world as they interact with other mechanisms. We established that these properties are an emergent phenomenon dependent on the components and relations that constitute the structure of the mechanism.

We then considered causality and drew a distinction between real causality and actual causality. The former concern the causal powers that mechanisms potentially have that may or may not be realized at any particular time, while the latter concern the actual events that do (or do not) occur through the mechanisms' interactions. In this section we considered two other issues – downward causation, where the whole causally affects its parts, and the idea that absences can have causal effects.

We then analysed the difference between powers and properties, concluding that they were indeed separate – a power can be the result of one or several properties, and a single property can create several powers. We next considered the distinction between the Real and the Actual based on the nature of events as opposed to enduring mechanisms. Events were found to be no more than identifiable changes to the disposition of mechanisms or systems. This led to a discussion of diachronic and synchronic causation which mirrors the earlier discussion of real and actual causation.

Finally, we moved away from the explanation of singular or particular events to the possibility of abstracting more general causal relations from series or patterns of events, what are known as demi-regularities. This was approached on the basis of Pearl's work which showed that it was possible, under certain circumstances, to deduce the existence of stable causal relationships from interrelated correlational data.

Notes

1 For a good introduction, see the entry in the *Stanford Encyclopedia of Philosophy* (http://plato.stanford.edu/).
2 Maturana and Varela distinguish between the *structure* of a system – its actual components and relations – and the *organization* of a system, which is a subset of the structure defining it as the type of system that it is.
3 What is called 'firstness' by Peirce (1931–1958, vol. 8, para. 329).
4 Called 'secondness' by Peirce (1931–1958, vol. 8, para. 329).
5 See a special issue of *Behavioral and Brain Sciences* (2003:26) dedicated to this issue.
6 The 'thing' or 'entity' may, of course, not be physical.
7 This distinction is the same as C.S. Peirce's (1931–1958, vol. 8, para. 329) phenomenological distinction between firstness and secondness (and then thirdness).
8 Others have brought in the relationship to Aristotle, e.g. Pratten (2009), Chakravartty (2008) and Witt (2008).

9 This tradition is actually not so recent: see for example Wright (1921), who first developed the idea of path coefficients in regression, and Haavelmo (1943), one of the founders of structural equation modelling.

10 Conditional independence is a probabilistic not a causal notion of itself. That is, it simply refers to whether knowledge of one variable affects the probability distribution of another variable. Unconditional independence between X and Y means that $P(X|Y)=P(X)$. Conditional independence between X and Y conditioned on Z means that $P(X|Y,Z)=P(X|Z)$. I.e. given that we know Z, then a knowledge of Y tells us nothing more about X. Z 'screens' X from Y. Note that such independencies are symmetrical: they apply equally to Y given X and so are non-directional.

11 Pearl does also call on one other principle – the stability principle. It is possible to generate minimal DAGS that depend on very particular combinations of parameter values. To rule these out it is assumed that the underlying mechanisms are stable with respect to changes in the parameter values.

6 Can social systems be autopoietic?

6.1 Introduction

In Chapter 5 we discussed the nature of emergence and causality from a primarily critical realist perspective but building on the ideas of mechanisms from Chapter 4. In the main, we were generally presuming physical or material systems and we mentioned that there were particular difficulties when considering whether these ideas could be applied to social systems. In this chapter we consider this question directly by asking whether or not social systems can be seen as autopoietic or self-producing.

We have already mentioned autopoiesis in several places (especially Chapters 1 and 3). It is one of the most significant developments in systems thinking in the last 40 years, and it certainly informs some of Bhaskar's ideas. But more than that, the concept of autopoiesis has been used by major social theorists such as Luhmann (1986; 1995) (explicitly) and Giddens (1979; 1984) (implicitly). It is therefore useful to take this as a unifying theme with which to review the many issues that are raised when considering the ontology of social systems.

In this chapter we will begin with an overview of autopoiesis and discuss why it is seen as potentially attractive for social theorists even though it originated as a biological theory of the nature of living systems. We will then consider Luhmann's social theory, which is explicitly based on his interpretation of autopoiesis. Then we will move on to consider Bhaskar's (1979) 'transformational model of social activity', as developed by Margaret Archer (1995; 2000a), in combination with Giddens' theory of structuration which, we will argue, has interesting commonalities.

6.2 The attractions and problems of social autopoiesis

The theory of autopoiesis, that is systems that are *self-producing* or *self-constructing*, was originally developed by Maturana and Varela (Maturana 1975b; Maturana and Varela 1980; Maturana and Varela 1987) to explain the particular nature of *living* as opposed to *non-living* entities. It was subsequently enlarged to encompass cognition and language, leading to what is known as second-order cybernetics (Maturana 1978a; Maturana 1978b; Varela *et al.*

1991).[1] However, as with earlier biological theories, many authors have tried to extend the domain of the theory to encompass social systems. Initial attempts (e.g. Beer 1975; Faucheux and Makridakis 1979; Robb 1989a; Zeleny and Hufford 1992) were sociologically crude and were greeted sceptically (Mingers 1989a; Mingers 1992a), but since then a major social theorist, Niklas Luhmann (1995), has made autopoiesis the basis of his social theory.

In this section we will consider critically the extent to which the theory of autopoiesis, *as originally defined*, can be applied to social systems – that is, whether social systems *are* autopoietic[2] – and, if it cannot, whether some weaker version might be appropriate. A major problem in this undertaking is that the nature of social systems and social structure is itself a highly debatable issue.[3] My strategy will be to take some of the major conceptualizations – Luhmann, Giddens, Bhaskar – and consider the extent to which autopoiesis is compatible with their theories.

6.2.1 The attractions of social autopoiesis

For reasons of space I shall not provide a detailed exposition of the theory of autopoiesis.[4] Suffice it to say that it is an ambitious theory aiming to explain both the nature of living systems and the biological foundations of human cognition and language in a way that recognizes the centrality of self-reference, recursion, and the process of (self-) observation. It does so in a very coherent and consistent manner, and has generated significant interest and debate in realms as diverse as law, family therapy, sociology, literature, and cognitive science (Mingers 1995b). What then are the attractions of autopoiesis as a social theory?

The very concept of a *self-producing* or *self-constructing* system has interesting and radical implications. Traditionally, systems theory itself has dealt with *open* systems that process or transform inputs into outputs.[5] Such a view can quite easily be applied to, say, an organization[6] – resources are taken in, undergo production processes, and result in products and services. Within sociology, systems theory was initially associated with Parsons (1951) and its main concern was seen as explaining pattern and order within society, rather than more dynamic processes of change and development. As such it was generally condemned for being overly functionalist, although there has been a revival in interest in Parsons' work: see for example Habermas (1987: Chapter 7). Buckley (1967; 1968) tried to address the problem of change with a much more dynamic view of systems. Societies were seen as complex adaptive systems which used internal feedback processes to change their structures so as to better survive in a turbulent and changing environment.[7]

There are, however, severe problems with this open-systems view. First, it gives primacy to the environment – it is the system that has to adapt itself to the environment. This seems to imply that it is the environment that somehow specifies or determines the structure of a social system. Yet, what exactly is the environment within which a social system, more especially a society, might exist? Is it the physical world, or other societies, or what? More generally, how would one draw a boundary to demarcate some well-defined social system that then interacts with an equally well-specified environment? Second, what could possibly be the inputs and outputs of such a system? Does it really make sense to conceptualize a society, or part of it such as a family, as a processor of inputs into outputs? More recently, ideas of complexity theory have been applied in sociology (Byrne 1998; Cilliers 2000; Walby 2007); to some extent Luhmann could be seen as part of the complexity movement.

An autopoietic system however is quite different. It does not transform inputs into outputs; instead it transforms *itself into itself*. What is meant by[8] this is that the outputs of the system, that which it produces, are its own internal components, and the inputs it uses are again its own components (it does always require some elements from the environment, and it does excrete waste). It is thus in a continual dynamic state of self-production.[9] The consequences of this are:

- A continual, circular process of production must be established that produces all that is necessary for continued autopoiesis. There is thus a deep closure of the system – it produces that which is necessary for it to continue to produce that which is necessary for it to continue to produce ... .
- Since the system produces itself, it gains a significant degree of autonomy – it depends less on other entities for its continual existence. At the same time, if it ever fails to produce that which is necessary then autopoiesis must break down and the entity will disintegrate. No functionalism is involved, however – the system either contingently maintains autopoiesis, or it does not.
- The theory distinguishes between the *structure* and the *organization* of a system. The structure (the actual components and their relations) may change dramatically over time, or may be realized in many ways, so long as the *organization* maintains its relations of self-production. It can be said to be *organizationally* closed but *structurally* open.
- The changes that can occur must allow autopoiesis to continue, and they are determined by the structure at each point in time. They are not determined by the environment, which can only trigger or select particular possibilities from those that the structure makes available (*structural determinism*).
- Systems can become structurally coupled to other systems, and to their environment, but this is a process of *mutual* specification (or co-evolutionary drift as Maturana would call it) rather than the adaptation of one system to another.
- The theory also incorporates cognition, language and observation in a coherent manner. It is anti-Cartesian in not separating cognition and action but emphasizing embodied cognition (Mingers 1995a; Mingers 2001b; Varela *et al.* 1991). It is non-representationalist in its view of language. And it embraces self-reference and recursion in generating a biological theory of the observer and, indeed, itself.

Given these characteristics of autopoietic systems, we can see several reasons why the theory might be attractive for sociologists:

- The distinction between organization and structure allows for radical change and development in a system without loss of its identity. This is very common in the social world, where we see many groupings – families, companies, religions, cultures and societies – that exhibit long-term stability and persistence despite enormous changes in their environment and in their own internal membership and structure.
- The organizational closure of the system means that we do not have to specify external inputs and outputs, nor do we have to see the system as functionally dependent on other systems. Its 'purpose' is simply its own continual self-production.
- The idea of structural determinism places the origin of change and development firmly within the system rather than from the environment, while the concept of structural coupling shows how, nevertheless, systems and their environments can mutually shape each other.
- The focus on self-production fits well with the ideas of Giddens (1984) and Bhaskar (1979; 1997), who both emphasize the way in which social structures are continually (re)produced and transformed through the social activities that they govern. It also resonates with Luhmann's (1982a) conception of a society functionally differentiated into subsystems, each essentially closed with respect to the others.
- The acceptance of self-reference and the ideas concerning language and observation also fit in well with the linguistic and communicative turn (Habermas 1979; Habermas 1984; Habermas 1987; Luhmann 1989a) in sociology and the greater recognition of the importance of the body (Featherstone *et al.* 1991; Grosz 1994; Shilling 1993; Synnott 1993; Turner 1984).
- The theory of autopoiesis resonates well with social constructivist (Gergen 1999) viewpoints – Maturana (1988) emphasizes the extent to which we 'bring forth' the world we experience through our own linguistic distinctions.

We can see that autopoiesis represents a major advance over previous systems theories in its sophistication and its potential for addressing many of the concerns of current social theory. However, there are major difficulties in simply applying the biological theory of autopoiesis to social systems, which will now be outlined.

6.2.2 The problems of social autopoiesis

While the idea of autopoietic organizations and social systems is very attractive, there are fundamental difficulties involved in such an application. If the concept is only to be used metaphorically, as Morgan (1986) suggests, to generate interesting insight then no great problems emerge – it is simply a matter of whether or not it is fruitful. To go beyond metaphor, however, and suggest that an organization or a social system *is* autopoietic raises significant ontological claims that are very difficult to substantiate (Brown 1988; Kickert 1993; Meynen 1992; Stokes 1990; Veld *et al.* 1991). This is already explicitly addressed in the work on autopoiesis at the physical level, where a clear distinction is drawn between the observer's descriptions and the operational autopoietic system. The problem is more acute with social systems. We, as observers, are trying to observe systems of which we are only a part, the constitution of which is still the subject of much debate. While we will not discuss these difficult philosophical problems in general, a number of aspects particular to autopoiesis need to be mentioned.

If it is to be claimed that social systems *are* autopoietic, then we must examine very carefully the specific criteria for autopoiesis and evaluate the extent to which they can be met by whatever we take to be social systems. There are three essential elements in the definition of autopoiesis:

1 Fundamentally, autopoiesis is concerned with processes of *production* – the production of those components which themselves constitute the system. It is therefore essential to identify clearly what are the components of an autopoietic social system, and what are its processes of production.
2 The autopoietic organization is constituted in terms of temporal and spatial relations, and the components involved must create a boundary defining the entity as a unity – that is, a whole interacting with its environment. In the case of social systems, is it possible to identify clear demarcations or boundaries that are constructed and maintained by the system?
3 The concept of the autopoietic organization specifies nothing beyond processes of self-production. It does not specify particular structural properties or components. It is thus so abstract or general that in principle it should be applicable to systems of any kind. The concept should not, therefore, need to be modified to deal with social systems. If it is so modified, can we still use the term 'autopoietic'?

In applying these ideas strictly, there are obvious problems. Is it right to characterize social systems as essentially processes of production and, if it is, what exactly is it that they are producing? If human beings are taken as the components of social systems, then it is clear that they are not produced by such systems but by other physical and biological processes. If we do not take humans as components, then what are the components of social systems? The emphasis on physical space and a self-defined boundary consisting of components produced by the system is also problematic. While space is a dimension of social interaction, it does not seem possible to sustain the central idea of a boundary between those components which are both produced by and participate in production, and those which are not. Again, taking people as components, they can choose to belong or not belong to particular groups or networks, and will be members of many at any time. What is it then that could constitute the boundaries of such a system? What might be its environment? And how can it be said that such a social system can act as a unity or whole – surely it is only individual people who act?

Overall, it seems difficult to sustain the idea that social systems are autopoietic, at least in strict accordance with the formal definition. However it is possible that a more generalized version such as Varela's (1979) idea of *organizational closure*, which does not specify processes of *production*, could be fruitfully applied. A more radical approach is to apply autopoiesis not to physical systems (such as groups of people) but to concepts or ideas or rules. Maturana defines a unity as 'an entity, concrete *or conceptual*, defined by an operation of distinction' (Maturana 1975a, my emphasis), and thus opens the possibility of a non-physical autopoietic system. Such a system might consist of concepts, or descriptions, or rules, or communications, which interact and self-produce.

6.3 Society as a system of autopoietic communication

We move now to consider the work of Niklas Luhmann who has, in fact, embraced autopoietic theory whole-heartedly and put it at the centre of his systemic social theory.[10] Luhmann's work up to autopoiesis is well covered in *The Differentiation of Society* (1982a), while his major development of autopoiesis is *Soziale Systeme* (1984; 1995 translation). Much of his theoretical writing is extremely abstract, but two translated books apply it to ecological problems (Luhmann 1989a) and the mass media (Luhmann 2000b).[11]

Luhmann's starting point is the idea that modern society[12] has become a functionally differentiated one. Societies are faced by an environment (all that is not society, not just the physical environment) which is inevitably more complex than they are. Over time, societies have become increasing complex themselves in response: that is, increasingly internally differentiated. This differentiation has taken many different forms (Luhmann 1982a, pp. 232ff.): for example, segmentation – the generation of many, essentially identical, subsystems such as villages in the Middle Ages; and stratification, where society splits into unequal subsystems forming a hierarchy such as capital and labour. Modern society can be seen as a development from stratified to functional differentiation. Subsystems become established in terms of the particular tasks that they carry out – for example, the economy, politics, law, science, the mass media, education, and religion. These subsystems become highly autonomous, distinguishing themselves from their environments self-referentially. Society no longer has a centre or controlling subsystem, but becomes the indeterminate outcome of the interactions between these independent but interdependent domains.

Luhmann brings in autopoiesis by arguing that all these subsystems, and society itself, are autopoietic unities and are thus organizationally closed and self-referring. In doing this, he recognizes the problems in defining social autopoiesis, in particular the exact nature of the components and the processes of their production (Luhmann 1986, p. 172). He accepts that social systems do not consist of, or produce, the (physical) people who participate in them.[13] So in what sense can they be autopoietic? His answer is similar to Varela's in suggesting that there can be closed, self-referential systems which do not have physical production as their mode of operation. These include both *social* systems and *psychic* systems (human consciousness). He differs from Varela in that he calls all such systems autopoietic, whereas Varela restricts that term to living physical systems.

6.3.1 Autopoiesis as the production of communication

So what are the basic elements of social systems which continually produce themselves? Not conscious thoughts, nor behaviour or actions, nor even language, but *communications* or, rather, *communicative events.*[14] 'Social systems use communications as their particular mode of autopoietic reproduction. Their elements are communications which are recursively produced and reproduced by a network of communications and which cannot exist outside such a network' (Luhmann 1986: 174). Each subsystem defines for itself what is and is not a communication for it, and then consists of networks of particular communications which always refer to previous communications, and lead on to other ones. Society as a whole encompasses all of the communications of its subsystems as well as the more general communications of the lifeworld.

It is important to understand what Luhmann means by 'communication', since he uses the term in a very specific sense. He stresses that it is not what we might normally mean by a communicative act such as a statement or utterance by a particular person. He characterizes a communication as an event consisting of three indissoluble elements – *information, utterance*, and *understanding* which can enable further communicative operations to occur (Luhmann 1995: 137). Each of these elements is said to be a selection from a range of possibilities. It is the operation of the autopoietic system which defines and makes the selections. Broadly speaking, *information* is what the message is about, *utterance* is the form in which it is produced together with the intentions of its sender, and *understanding* is the meaning that it generates (which can include misunderstanding) in the receiver. There must be at least two parties involved, the communicator and the receiver, but these do not have to be specific individuals.

Communication is not the simple sending of a message – it cannot be said to have occurred until the receiver has received and understood something, even if not what was intended. Indeed, the nature of the communication remains undefined until it has been interpreted by the other. Nor can it be understood as the *transmission* of some *thing* (information) from one person to another. The utterance is a selection, a skilled performance chosen to provoke or trigger a reaction in the receiver. But it can never *determine* what the reaction will be, for this too is a complex selection based on the receiver's own cognitive state.

All these elements are generated or co-produced together as a unity, and this event allows the possibility of further communications. This happens through a fourth selection, by the receiver: the acceptance or rejection of the communication's meaning (Luhmann 1995: 147). This is distinct from understanding. Any communication generates meaning, whether intended or not. The fourth selection is the link to action – does the receiver respond in some way to the communication, perhaps to question or disagree, or does the receiver fail to respond and thereby terminate the communicative sequence? It is important to stress that all three aspects are distinctions made by the (sub)-system itself, not by an outside observer. The system determines what, for it, is information; how it may be embodied; and how it may be interpreted. In doing this it draws its own distinction as to what belongs to the system and what does not – this is the closure of autopoietic systems.

We can visualize the whole system as an on-going network of interacting and self-referring communications of different types, and see how they can be separated from the particular people involved. The people will come and go, and their individual subjective motivations will disappear, but the communicative dynamic will remain.

Having looked at the structure of communication, we now ask: how are the dynamics of autopoiesis constituted? Essentially it is a network through time of communications referring to other and past communications and leading to new ones. However, it is quite a different form of production from physical autopoiesis, for communications are events. They occur (or at least are completed) at a point in time and then disappear. They may leave traces – memories, papers, films – but these are not the events themselves. What is vital is the generation of the next (communicative) event, for without this autopoiesis stops; and that this event is different from the previous one. So communicative autopoiesis is a production not of structure or pattern or repetition but of networks of differentiated events. In terms of the distinctions discussed in Chapter 5, communication networks are inherently diachronic. A question to be addressed is: what are the synchronic mechanisms that generate them?

Looking specifically at the relationships between communications, Luhmann (1986: 175) says that they will either be hetero-referential or auto-referential. A later communication distinguishes between the information and the utterance of an earlier communication. It then either concerns the information – questioning it, denying it, or enlarging on it (hetero-) – or it concerns the utterance, asking how it was said, why it was said, or who said it (auto-). In each case, the later communication makes its own particular distinctions (or selections) among these varied possibilities.

We can see here the relationship of communication to meaning (Luhmann 1985b; Luhmann 1990a). Events (especially communications) refer to and are related to many other events and possibilities. The production of communication is precisely this selection from the manifold possibilities – distinguishing what is by what it is not. It is these related events and possibilities which constitute meaning. Meaning is the openness of all possibilities: all the relations, distinctions and denials that could be generated – a very Husserlian construction. It is that which provides newness and difference between communications. On the other hand, a particular communication closes this off – it fixes one possibility in order that something might actually happen. Autopoietic communication can thus be seen as meaning-processing (Luhmann 1989a: 17), generating distinctions to convert the open field of meaning into the particular information/utterances which thereby constitute a society.

6.3.2 The autopoiesis of society

> A social system comes into being whenever an autopoietic connection of communications occurs and distinguishes itself against an environment by restricting the appropriate communications. Accordingly, social systems are not comprised of persons and actions but of communications.
>
> (Luhmann 1989b: 145)

Society differentiates itself into subsystems each of which is an autopoietic network of recursive communications. Society itself is also autopoietic, consisting of all these communications plus all others not specifically involved in subsystems – the communications of the lifeworld (Luhmann 1989b).[15] As such, it distinguishes itself from its environment – that which is not communication. Thus, not only the physical environment but also people and their consciousnesses are in the social system's environment. Only thoughts can generate thoughts and equally only communications can generate communications. Society is a closed system in that it cannot communicate directly with its environment, since the environment, by definition, does not communicate. Events happen in the physical world (e.g. pollution) but this does not affect society until it becomes the subject of a communication – society cannot communicate with but only about its environment, according to its capacities for information processing. This does not mean that society is totally isolated – it is like physical autopoiesis, organizationally closed but interactively open. The environment (especially people) can trigger or irritate society and society may then generate a communication, but the nature and form of that communication will be determined by society or by a particular subsystem, not by the environmental disturbance.

In a similar way, the subsystems also distinguish themselves within society and specify their own boundaries. They too form closed networks of communications – each one only being able to process or deal with communications of its own type. Luhmann analyses their workings in terms of codes and programs. He argues (Luhmann 1989b: 36ff.) that each subsystem utilizes a particular binary code representing the good/bad, positive/negative for that subsystem. For example, the code for the law is legal/illegal, for the economy to pay/not pay, for science truth/falsity, for politics the holding/not holding of office, for the mass media information/non-information. The code provides the basic guidance for a subsystem, for without it the self-referential operations would be entirely undetermined. The code itself is just the particular categories, and it therefore requires some means or criteria for assigning events to a category. This is the program – the rules for coding. This separation is one way in which variety can be increased, since it allows the program to be changeable even though the code is not. For example, the historical development from natural to positive law involves a shift from defining legality in terms of religious or natural criteria which were unchangeable to criteria which are defined by society and are thus open to change.

For subsystems, the other functional subsystems exist as part of their environment. There are much greater interactions and dependencies between subsystems than between society and its environment. The subsystems have become autonomous and independent, but at the same time more interdependent since they rely on the existence of the other subsystems to carry out particular functions. Interactions between subsystems are reasonably well defined – legal communications can give rise to economic ones which in turn trigger political ones. When a subsystem is triggered by its environment and generates a communication about a particular matter, this is called resonance.[16]

There are several types of (structural) couplings. First, the coupling of communication (i.e. society) to sense-systems – that is, individual consciousnesses – but not, Luhmann argues, to the general physical world. Then there is the coupling of subsystems to society itself. Indeed, this coupling is most close, since the subsystems are not something other than society but part of its very constitution. Nevertheless, they do distinguish themselves through their operation. Finally, there is structural coupling between subsystems, and here Luhmann details a few specific combinations. The economic and legal subsystems are mainly linked through the law of property and contract, and politics and the legal system by constitutional law. Events in these particular fields generate communications in both the connected subsystems which then become part of the subsystems' autopoietic operation.

We are thus left with a view of society very different from the traditional one. Society is essentially centreless – there is no core or fundamental division driving it, and there is no privileged position from which a rational overall view can be developed. Instead we have self-defined autonomous subsystems in a constant process of renewal and redefinition, locked together in a fragile balancing act, resonating among themselves but relatively unresponsive to society's external environment.

6.3.3 Luhmann's autopoiesis – evaluation

The question now is, to what extent can we accept Luhmann's social theory as genuinely and correctly embodying the underlying theory of autopoiesis as set out above? The conditions to be met were articulated above: to be able to clearly specify the components and the concomitant processes of production of those components; and to be able to identify a clearly demarcated boundary so that the system can be said to act as an organizationally closed unity and to produce itself as a whole.

Certainly Luhmann's work represents a bold attempt to theorize an autopoietic unity in the non-physical domain. It defines the basic components of such a system – in this case communications – and holds consistently to this without confusing domains by, for example, including people within the system. The nature of production is shifted to a production of events rather than of material components. Finally, the circular and self-defining nature of the production network is brought out well, as is the combination of organizational closure and interactive openness. Examining the system in more detail, however, shows up several problems.

First, let us consider the notion of components and their production. That the components are *events* does not seem a particular difficulty – ultimately it is just a matter of time scale. Given long enough, all production processes become events since the produced component will exist and then disintegrate. Equally, with a short enough time horizon all events themselves become processes as we observe their unfolding.

However, I would argue that there is a problem with the notion of *production*, specifically with the claim that it is communications (rather than people) that produce further communications. Put another way, there is little attempt to show how societal communication, as an independent phenomenal domain, emerges from the interactions of the human beings who ultimately underpin it. Without human activity there would be no communication. Maturana is always careful to show how new domains arise out of the interactions of observers, but with Luhmann the observer is lost completely in favour of the observation. This is an important lacuna with a number of concomitant problems, some of which are also identified by Habermas (1985). How do communications actually occur? It is one thing to say analytically that communications generate communications, but operationally they require people to undertake specific actions and make specific choices. Is not the claim that 'communication produces communication' stretching the notion of production too far? One communication may stimulate another but surely it does not produce or generate it.[17] How does this interaction occur? What factors affect the selections that are actually made? In general, what is the relationship between the psychic systems of individual consciousness and the social system of communication?

Luhmann does consider this analytically with his concept of *interpenetration*. This describes the way in which something can be an element in two systems at the same time. Thus an action (e.g. an utterance) is part of the psychic system of human activity. At the same time it can be used, *as an element*, within a social system of communication, but it will not be the 'same' element in the two systems – it will have different functions. '[I]nterpenetrating systems converge in individual elements – that is, they use the same ones – *but they give each of them a different selectivity and connectivity, different pasts and futures*' (Luhmann 1995: 215, original emphasis). While this would seem to be true descriptively, it does not provide sufficient explanation of the complex interaction between the two levels. Put in the terms of Chapter 5, there is only diachronic causality in Luhmann's theory; there is little consideration of synchronic causality in terms of the events being generated by underlying mechanisms.

Luhmann's theory would seem to rely on his concept of meaning as the link between the two. A communication opens up possibilities through its meaning to people whose selections then generate new communications. However, this appears to be a very individualistic analysis. Little attention is paid to the constituting of these subjects as subjects and the major role of language in this. In Maturana's terms, Luhmann ignores the importance of languaging and consensuality which provides an already existing, a priori, structure generating an intersubjective domain of interactions. Such an approach would provide a general bedrock in the lifeworld out of which the particular specialized communicative subsystems can be seen to arise.

Second, how does Luhmann's theory deal with the question of boundaries? At first sight it looks promising. In defining components as communications it clearly distinguishes them from that which is not communication, e.g. the physical or psychological domains. It then specifies functional subsystems within society as a whole that demarcate themselves by distinguishing their own communications. While this does not actually involve boundary components, it does (to the extent that it is successful) generate a clear cleavage in the communicational space.

The question is whether, empirically, subsystems can be properly described in such a pure fashion. In the economic sphere, for example, it may be that the ultimate operations underlying it are monetary payments, but if we seek to explain particular happenings we immediately find that social, political and legal factors are at work. This is because it is people who make economic decisions – to buy, or sell, or invest, or lend – and people form a nexus between all the different subsystems. Their decisions are affected by their expectations, which are conditioned by communications from other domains. Moreover, communications can often be said to belong to more than one domain. For example, signing a loan agreement both generates a transfer of money and establishes legal obligations. Awarding a research grant is both a communication about academic status and a payment. As another example, Luhmann characterizes organizations as recursive networks of communications about *decisions* and only decisions (Luhmann 2000a). Yet this seems an incredibly reductionist view of the rich complex of social interactions within an organization.

To summarize, this section has developed the following conclusions concerning Luhmann's application of autopoiesis to social systems. In terms of components and processes of production, we can take the components of such a system to be communicative events, where communication consists of information, utterance, and understanding. However, there are problems in accepting that communications are *produced by* other communications alone rather than by people within social interaction. This is part of the problem of the totalizing nature of the definition of society as communications and only communications, leaving the mutual interaction between people and society under-theorized.

In terms of organizational closure, the idea that society, and its subsystems, define their own boundaries through communications that do or do not belong to them has some potential; it is difficult, however, to accept empirically that subsystems can be adequately characterized in such a pure and separable manner.

It could be said that theoretical purity (in applying the theory of autopoiesis) is gained at the expense of an incredibly abstract and reductive view of the social world. Society, and the rich processes of social interaction between real people, become marginalized in favour of almost entirely disembodied communicative mechanisms. It is as though an analysis of e-mail usage considered only of the flow of electronic messages around computer networks with no thought for the content of the messages or the motivations of the people involved. Yes, that is one level of analysis, but surely not the only one.

6.4 Structuration theory and critical realism

Giddens' (1979; 1984) concept of structuration is one of the main developments in recent social theory, and it does have definite resonances with autopoiesis:

- Giddens' concern with the continual, recursive (re)production of social structure through time is clearly linked to the idea of self-producing systems (Mendoza 1997; Mingers 1996a). In places, the idea of circular self-production is very clear:[18]

  > 'By its recursive nature I mean that the structured properties of social activity – via the duality of structure – are constantly recreated out of the very resources which constitute them.'
  >
  > (Giddens 1984, p. xxiii)

  > 'Human social activities, like some self-reproducing items in nature, are recursive. That is to say, … in and through their activities agents reproduce the conditions that make these activities possible.'
  >
  > (Giddens 1984, p. 2)

  > '[A]ll social life has a recursive quality to it, derived from the fact that actors reproduce the conditions of their social existence by means of the very activities that – in contexts of time-space – constitute that existence.'
  >
  > (Giddens 1987, p. 221)

- Both theories (autopoiesis and structuration) emphasize that explanations should be non-functionalist and non-teleological.
- Both distinguish between that which is observable, having space-time existence, and that which is not but is still implicated in the constitution of a system (*structure/organization* for Maturana, *system/structure* for Giddens).
- Both take an essentially relational view of social systems and identify the same three sets of relations: constitution/space, order/time and specification/paradigmatic.[19]

At the same time, a related conception of social systems (the transformational model of social activity – TMSA), is becoming influential. This was first articulated by Bhaskar (1979) as part of a comprehensive post-positivist, but realist, philosophy of natural and social science, and further developed in later works (Archer *et al.* 1998; Bhaskar 1993, 1994; Harvey 2002). At first sight this appears compatible with structuration theory with talk of the duality of structure, of the continual reproduction and transformation of society, and the notion that social structure only exists in and through the activities it governs. However, there has been considerable debate about the degree of similarity as will be discussed below. Even more pertinent, however, is that Bhaskar specifically uses the term 'autopoiesis' to describe fundamental aspects of the TMSA (as was pointed out in Section 3.4.3), although nowhere does he actually reference Maturana and Varela's writing.

Thus there is clear *prima facie* evidence that the theory of autopoiesis has been influential for both Giddens and Bhaskar.

In the next section I will first briefly cover the Giddens–Bhaskar debate and argue that the two are in fact potentially compatible, and then consider the extent to which autopoiesis may be involved in this conceptualization. Before that, it is necessary to mention a fundamental objection to the whole notion of social autopoiesis: that social systems or social structures do not exist in a causally efficacious sense at all.

The basic contours of this debate are as old as sociology itself. They used to be defined in terms of individualism versus collectivism and now tend to be discussed through the agency/structure distinction. Critical realism (strongly) and structuration (less strongly) maintain that social structure(s) exist, and have causal effects, over and above the actions of individual people. Upward conflationists, to use Archer's term (Archer 1995) (downward reductionists as I would call them), deny causal reality to social structure, which they see as simply an effect or epiphenomenon of individual social interaction. This case has been argued strongly in terms of the primacy of hermeneutics by King (1999a, 1999b, 2000) against both Giddens and Bhaskar, and has been rebutted by Archer (2000b). There has been a related debate, couched in terms of whether or not social structure can be causally efficacious, between Charles Varela and Harré (Varela 1999, 2002; Varela and Harré 1996) and Lewis (2000) and Bhaskar (2002a).

I do not wish to enter this debate directly, but it is clear that any attribution of autopoiesis to the social world must presume the existence and efficacy of social structure. To accept the opposite would immediately foreclose the possibility of social autopoiesis. So, for the purposes of this chapter I will simply take as given that the characterization of the social world is not exhausted by the individual's meanings and actions, while recognizing that this is actually an on-going debate.[20]

6.4.1 Giddens and Bhaskar: a distinction without a difference?

It is inevitable within social science that particular authors will use certain basic terms such as 'society', 'social structure', or 'social system' in different ways.

There is then much debate about these terminological differences that may obscure the similarities of the underlying conception or model. Certainly there are differences of substance and emphasis between Bhaskar and Giddens, not least because they approach their common object from different directions – philosophy and sociology respectively; but within the wide realm of social theory as a whole they seem to me to occupy essentially the same niche.[21]

We have seen that Bhaskar's central realist proposition is that, in both the natural and social worlds, there exist structures or mechanisms with particular powers or tendencies (the domain of the Real). These interact in such a way that actual events occur (or do not occur) (the domain of the Actual) and some of these are observed or experienced by people (the Empirical) as part of the process of developing knowledge. These mechanisms or systems do not have to be physical, or even observable – the criterion for existence is the non-empiricist one of causal efficacy. This means that concepts, ideas, rules, and practices, for instance, are no less real for being unobservable (Bhaskar 1997).

For Bhaskar, society exists as an object in its own right, emergent from but separate from people and their activities, and with its own properties.[22] Society always exists before individuals, who do not therefore create it but only transform or (re)produce it. Nevertheless, society is *necessary* for social activity and it only *exists* in virtue of that activity. Society therefore conditions social activity and is either maintained or changed as an outcome of that activity (*the duality of structure*). Equally, human action (praxis) is both a conscious production, i.e. intentional bringing about of purposes, and an unconscious (*re*)production of society (*the duality of praxis*).

> At the heart of this idea is the conception of human agency or praxis as transformative negation of the given . . . ; and at the same time as both enabled and constrained by and reproductive or transformative of the very conditions of this praxis, so that these conditions are activity-dependent or auto-poietic.
>
> (Bhaskar 1994: 93)

Society is said to be an 'ensemble' of structures, practices and conventions, where structures are relatively enduring generative mechanisms that govern social activities. Being more specific, Bhaskar suggests that there must be a linking mechanism between human action and social structure and that this mediating system consists of *position-practices*: that is, combinations of roles that can be filled and practices that are then engaged in. It is important to note that position-practices are relational – they develop to form a system in relation to each other and this is separate from any network of relations between those who happen to occupy them. While emphasizing the ontological reality of social structures, Bhaskar recognizes that they have significantly different properties from physical objects. In particular:

- Social structures do not *exist* independently of the activities they govern.
- Social structures cannot be *empirically identified* except through activities.

- Social structure is not independent of actors' *conceptions* of their activity.
- Social structures are *relative* to particular times and cultures.

Despite these differences they are still suitable subjects for scientific theorizing, even if this lead to particular epistemological difficulties (Bhaskar 1979).

Structuration theory also has a core distinction at its heart – that between social *structure* and social *system*. Taking first social *structure*, this describes not empirically observable patterns or regularities, as in positivism, but underlying sets of rules that generate the observed regularities, more akin to structuralism (and not far from the idea of causal mechanisms).

> Structure, as recursively organised sets of rules and resources, is out of time and space save in its instantiations and co-ordination as memory traces, and is marked by an 'absence of the subject'. The social systems in which structure is recursively implicated, on the contrary, comprise the situated activities of human agents, reproduced across time and space.
>
> (Giddens 1984, p. 25)

Structure is thus seen as similar to a code or set of rules that governs possible selections of social action. It is constituted as an 'absent set of differences' that is not empirically observable as such, but is only exhibited in particular social interactions. Its existence is said to be 'virtual'. In fact, structure should really be seen as the structuring properties of social systems; these properties can be understood as rules and resources, recursively implicated in the reproduction of social systems. In many ways this is a similar conception to Luhmann's, although covering more than simply communication.

Social *systems*, by contrast to structure, do exist in time-space, and consist of observable activities and practices. The regularities that we can observe in social systems occur both spatially and over time, and this observable patterning and interdependence is brought about and sustained through the virtual (unobservable) structure governing their activity.

The relationship between system and structure is provided by the concept of *structuration*, a twofold process which Giddens sometimes refers to as the duality of structure. First, structure organizes the practices that constitute a social system – actors draw on the structural rules and resources in the production of interaction. But, second, it is precisely and only these interactions that reconstitute (and possibly transform) the structure. 'The structural properties of social systems are both the medium and the outcome of the practices that constitute those systems' (Giddens 1979, p. 69). This is the central kernel where both the TMSA and structuration appear to be an embodiment of autopoiesis.

6.4.2 TMSA versus structuration

In the beginning there appeared to be clear resonances between structuration theory and Bhaskar's early social theory, as even Archer (1995, p. 147) accepts.[23]

The following could easily be describing structuration theory: 'On this model, unintended consequences and unacknowledged conditions may limit the actor's understanding of their social world, while unacknowledged (unconscious) motivation and tacit skills may limit his or her understanding of himself or herself' (Bhaskar 1986: 125).[24] However, on deeper inspection it became clear that there were in fact substantive differences. Bhaskar wrote: 'This [analytically discrete moments of social interaction] is a feature which, as Margaret Archer has convincingly demonstrated, distinguishes it [TMSA] from structuration, or more generally any 'central conflation' theory' (Bhaskar 1993, p. 160, my insertions). These differences revolve around the ontological status of social structure in the two theories. For Bhaskar, there is a *dualism* of two distinct entities – people and their social activity on the one hand, and the social structure(s) that emerge from and also enable and constrain such activity on the other. Both are equally real. For Giddens, there is a *duality* between observable social systems and their unobservable, virtual structural properties.

Several writers from the critical realist camp have been deeply critical of structuration theory, for instance Archer (1990, 1995), Layder (1985, 1987, 1989), Craib (1992), Thompson (1989) and Porpora (1989). The fundamental claim, made by each author in different ways, is that Giddens does not give sufficient ontological independence to social *structure* (or system); that he essentially treats agency and structure as though they are inseparable, two sides of the same coin, with centrality being afforded to the encompassing notion of social practices. This means that Giddens remains too much on the subjectivist side of the fence, refusing to accept the leap into an objective, constraining social structure.

I accept that there is force to this argument. Giddens adopts a perspective that puts much greater emphasis on the knowledgeable activities of agents and does not recognize a separately existing social structure in the way that Bhaskar does. However, I will argue that the anti-Giddens camp go too far and erect something of a straw man in their characterization of Giddens, who would not hold some of the extreme positions that they impose on him. This argument then allows us to consider a possible synthesis of the two models.

We begin with Archer (1995), whose position can be summarized as follows:

> A realist ontology which regards structural and cultural systems as emergent entities is at variance with the Elisionists' [Giddens *et al.*] view which holds, (a) that such properties possess a 'virtual existence' only until, (b) they are 'instantiated' by actors, which (c) means these properties are neither fully real nor examinable except in conjunction with the agents who instantiate them.
>
> (Archer 1996: 692, my insertion)

This, according to Archer, has several consequences:

- Both elements, agency and structure, are denied autonomy and their own separate properties since both are subsumed under social practices. This has the effect of flattening ontological strata, losing both that of social system and that of psychological individual.

- This means that we cannot investigate each as a separate entity, except in the limited sense given by Giddens' methodological bracketing; nor can we consider the ways in which agency and structure, as independent entities, might causally interact with each other.
- The time dimension is lost. Since structure and agency are simply different reflections of the *same process*, they must be simultaneous. We cannot conceptualize how structure at time t conditions activity at t+1 which then transforms or reproduces structure at t+2. Archer suggests that Giddens moves from the obvious 'no people: no society' to the questionable 'this society; because of these people here present' (Archer 1995, p. 141).
- This also makes it difficult to understand under what conditions social activity will change rather than simply reproduce the pre-existing conditions. This can be put another way in terms of the difference between *social* and *system* integration. For Giddens, the distinction is primarily one of scale – face-to-face relations as opposed to relations between collectivities at a distance. Archer argues that there cannot therefore be a disjunction between the two, whereas a separation of the two would allow different degrees of integration/conflict in the two domains so that, for example, social conflict may or may not result in systems change.

Porpora (1998b) examines four different concepts of social structure: stable patterns of aggregate behaviour (e.g. Homans or Collins); lawlike regularities among social facts (e.g. Durkheim or Blau); systems of relations among social positions (Bhaskar); and virtual rules and resources (Giddens). He argues that Giddens is a realist in accepting that structural rules and resources do causally affect social activity, but is not enough of a realist to also grant causal efficacy to the 'objective' social relations to be found in Giddens' social system. Rules and resources are important, but are ultimately subjective (or intersubjective) in necessitating some degree of at least tacit understanding and knowledge on behalf of actors. In contrast, Porpora suggests that of more fundamental importance are the material, objective social relationships such as the distribution of income, the division of labour, and job opportunities that act as external constraints on individuals. The heart of the disagreement is that 'Giddens gives analytical priority to rules and in fact denies that the relationships of a social system have any causal properties independent of the rule-following activities of human actors' (Porpora 1998b: 350), whereas Porpora maintains that social relations do constrain in a way that is independent of the actor's knowledge of them.

Layder's (1985; 1987; 1989) critique seems to rest on a rather partial reading of Giddens' work. For instance, one of Layder's main arguments is that the idea that social system and social structure must always be instantiated through social activity loses an essential distinction between such activity and pre- and post-existing system/structure. In support of his view he quotes (Layder 1987, p. 34)

Giddens' assertion that 'social systems only exist in so far as they are continually created and recreated in every encounter as the active accomplishment of subjects'. But Giddens is saying something rather different: not that social systems only exist at all in the moments of their instantiation but that they will no longer exist if they are not continually re-enacted. He is simply making the point that particular practices will only remain in existence if they are, in fact, practised. Indeed, the use of the term '*re*created' in the quote clearly acknowledges the fact that there is something already existing, which is recreated or reconstructed through social activity.

A similar misinterpretation occurs in Layder's discussion of the extent to which social structure can constrain action. He construes Giddens as saying that constraints can only be identified with that which is internal to a particular episode of social interaction and indeed ultimately with the psychological motivations of the actors involved. What Giddens actually says, as quoted by Layder (1987, p. 39), is 'Structural constraints do not operate independently of the motives and reasons that agents have for what they do.... The only moving objects in human social relations are individual agents who employ resources to make things happen, intentionally or otherwise.' Layder then makes the illicit equivalence that 'The word "operate" doubles for the word "exist"' (Layder 1987, p. 40).

Again, I would argue that Giddens is saying something significantly different and that 'operate' does not in fact equal 'exist'. Giddens' point, and it is a very fundamental one that I believe is accepted by Bhaskar and Archer, is that only people can actually undertake social activities. Systems, structures, practices or whatever do not, of themselves, act – only people can do that. So structural constraints can only have effects (operate) by affecting people, and in particular by shaping their motives and reasons for action. This does not mean that such constraints do not *exist* independently of and prior to the activities of particular individuals. It simply means that the powers of the constraint are not actualized (to use a Bhaskarian term) except through people. Nor does it mean that the actors involved have full transparency over the process. Giddens accepts that there are both unacknowledged conditions of action and unknown consequences of action. So it is quite possible for constraints to determine aspects of the contexts within which people find themselves and thereby shape the choices made without those involved being fully aware of it.

With regard to the fundamental question of ontology, it seems to me that there is a substantive difference between Bhaskar and Archer's dualist model and Giddens' dualism at least in so far as social *structure* is conceptualized, although both see social structure as only *existing* and *observable* through social activity, and inevitably dependent (to some extent) on actors' knowledge of what they are doing. But I do not accept Archer's (and Bryant's, 1995, p. 97) view that Giddens cannot therefore be seen as a realist because of the *virtual* nature of his concept of structure. I would argue that this is a mistaken interpretation of the term 'virtual'. Giddens uses this in contrast to those things that have space-time presence – that is, that happen at particular times and places. Virtual rules and resources do exist; they are *real*; they are as Giddens says 'generative' – they do have causal effects; but they endure and underlie the events that they enable. Indeed, the distinction is very close to that which Bhaskar makes between the domain of the Real (enduring mechanisms) and the domain of the Actual (particular events). Thus it is not virtual as opposed to real, but virtual as opposed to actual.

I do, however, think there is a substantive difference in the way Giddens and Bhaskar conceptualize the term 'structure', and this is the basis of much of the problem. Bhaskar takes a traditional view that out of the social activities of people a new entity emerges – *society*. This is said to consist of various *structures* – that is, relational systems of position-practices that govern, and are reproduced/transformed by, social activity. This is essentially the same as Giddens' social system,[25] consisting of practices which, when long-standing and widely spread, are termed 'institutions', a term also used by Bhaskar. What Giddens then does is to highlight a particular aspect of the mechanism whereby social systems govern activity and the activity reproduces the system – that is, rules and resources. Practices and institutions, which can be observed, must have rules[26] underlying them in order for the activities to occur, although these rules will not be observable save through the activities. In calling these rules and resources 'structures' Giddens recognized that he was moving away from the common usage of the term although he was not uncomfortable with its continued traditional use, as in 'class structure' (Giddens 1984: 19). It could be said that Bhaskar's usage of structures as 'generating mechanisms' is itself a new development.

I would suggest that the two views can be reconciled by using 'structure' in Bhaskar's sense and saying it consists of positions, practices, and the rules and resources that underlie them,[27] but then using much of Giddens' substantive theorizing about how such a complex and stratified structure interacts with praxis.[28] Although the objection might be raised that this would return from duality to a dualism, there are elements of structuration theory that seem to fit and to answer some of the more detailed points raised by Archer above.

First, Giddens recognizes that structures may be transformed, not simply reproduced. In fact he identifies four different mechanisms of social change (Giddens 1990b): *system reproduction* – the gradual and unintended drift of social practices; *system contradiction* – conflicts of interest within and between social systems; *reflexive appropriation* – conscious shaping of social systems, especially organizations; and *resource access* – changes generated by changing availabilities of resources. This implies that there is a degree of distanciation between structure and system – rules are not causally determinative but may be enacted in different ways, and the consequences of action, intended or not, may bring about structural change rather than reinforcement.[29]

Second, Giddens does recognize the temporal element in the structure/action relation, the idea that actors do not create *de novo* but always transform or reproduce something that exists before them. 'Human societies, or social systems, would plainly not exist without human agency. But it is not the case that actors create social systems: they reproduce or transform them, remaking what is already made in the continuity of praxis.' (Giddens 1984: 171.)[30] Further, in discussing the structuring of institutions he says that this 'raise[s] once more the problem of history, since the absent others include past generations whose time may be very different from that of those who are in some way influenced by residues of their activities' (Giddens 1984, p. 37). He even accepts that all social life, from the micro to the macro, is inevitably 'episodic' (Giddens 1984, p. 244): that is, it can be regarded in terms of sets of events having specifiable beginnings and ends during which significant changes to the social structure may occur. This all goes against Archer's assertion that structuration theory limits itself to the activities of presently existing people and is unable to recognize the effects of an already existing structure.

Third, when considering specific mechanisms by which social institutions are reproduced, we can see causal relations between system and structure. The concept of structure itself is stratified into different levels of abstraction (Giddens 1984: 185). The most abstract and enduring are 'structural principles', which underlie the organization of whole types of society – e.g. capitalist. At the next level are 'structures', which are particular sets of transformation relationships between elements within a society, e.g. the relations between commodities, money and capital.[31] Finally, there are 'structural properties' or 'elements of structuration', which are the most concrete, linking specific systemic occurrences with wider societal institutions. An example is the division of labour – a general structural property that is enacted within particular organizations. These are linked to dynamic processes of reproduction or change – what Giddens calls homeostasis and reflexive regulation[32] or circuits of reproduction (Giddens 1984: 190). An example is the poverty cycle of deprivation – poor schooling – poor jobs – deprivation – poor schooling. Clearly such causal loops can be seen as structural generating mechanisms the exercise of which results in particular, observable phenomena.

With regard to Porpora's criticism, I suggest that he is imposing a rather crude dichotomy onto what is actually a complex mix of known and unknown conditions of action. Giddens (1989) himself speaks of three ways in which action may be constrained – first, the material constraints of the body and the physical world (which can of course be changed through technology); second, constraints stemming from the direct application of some form of power or sanction (which can vary in intensity); and, third *structural* constraints imposed by the context of action of an individual.

In this latter case, Giddens recognizes constraints deriving from the pre-existing social situation and from the social relations in which actors find themselves: 'All structural properties of social systems have a similar "objectivity" *vis-à-vis* the individual agent. How far these are constraining qualities varies according to the context and nature of any given sequence of action.' (Giddens 1984: 177.) He goes on to accept the legitimacy of a sociological explanation in terms of social forces (such as technology) 'without reference to agents' reasons or intentions'. However, structural constraints are not causally determinative in the way that physical forces sometimes are, and they also differ in always being enabling of action as well as constraining it. He insists that ultimately all such constraints must work through individual (or groups of) actors by restricting the range of choices available in particular situations – the greater the degree of constraint, the fewer options available. Thus, there is in principle always some degree of choice even when actors feel they have but one course of action. This does not imply that actors are always (or ever) fully aware of many of the conditions or consequences of their activity. There are limits here in terms of both unconscious motivations and unknown conditions of action. 'It is equally important to avoid tumbling into the opposing error of hermeneutic approaches and of various versions of phenomenology which tend to regard society as the plastic creation of human subjects' (Giddens 1984, p. 26).

To summarize this section, I have tried to show that Giddens' and Bhaskar's conceptions of society and social structure, while different, can usefully be synthesized. Social structures, consisting of position-practices, rules, and resources, are generating mechanisms that, through their complex interactions, enable and constrain observable social activity which in turn reproduces and transforms these structures. Society is then a particular combination of both praxis and structure that is historically and temporally located.

6.4.3 Autopoiesis and social structure

There are two questions to be answered in this section: is it possible to apply autopoiesis to the social theory outlined above? And, if it is, what benefits would this bring?

We can see elements of circularity, self-reference, and production in the above description. First and foremost, as illustrated by the quotations at the beginning of Section 6.4, is the mutual dependence of praxis and structure. Social activity could not occur without a pre-existing structure, but the structure itself is only produced and reproduced through the activity. In a very general sense this must be seen as self-production – to take Bhaskar's two dualities, structure continually produces itself through its enactment in praxis, while praxis continually produces its own pre-conditions through its crystallization in structure. Going below this generality, we can see that Giddens especially has identified many specific causal loops or circuits of reproduction that can be seen as akin to the sorts of chains of chemical reactions which occur within cells. However, is this enough for us to accept social autopoiesis? The conditions to be met were articulated above – what are the components and what are the processes of production of those components? And can we identify a clearly demarcated boundary so that the system can be said to act as an organizationally closed unity and to produce itself as a whole?

With regard to the components, it seems clear that they cannot be the actors themselves for they are the result of systems of biological production. I also do not think that it could be their actions or activities as such, for while these may be conditioned by social structure they are surely not, in general, produced by it, in the same way that the structure of language enables and constrains what *can* be said, but not what *is* said. Peoples' actual actions surely result from their own stratified and historically situated selves, albeit reacting to a particular social context or situation, and expressing themselves through legitimized forms of behaviour. This only leaves the elements of social structure – rules, positions, practices, etc. – as potential components for social autopoiesis. This fits in quite well with the paradigmatic example of non-physical autopoiesis, the game of Nomic,[33] discussed in Mingers (2002). Here, it was specifically the rules of the game that were the self-producing system rather than the players or their actual moves.

If these are the components, what then are the processes of production that generate them? First, we need to consider what sense can be given to the term 'produce' here; this is one of the problems Varela himself highlights – instantiating the concept of production within the realm of social systems.

> In order to say that a system is autopoietic, the production of components in some space has to be exhibited; further, the term production has to make sense in some domain of discourse. Frankly, I do not see how the definition of autopoiesis can be directly transposed to a variety of other situations, social systems for example.
>
> (Varela 1981b: 38)

When applied to biological systems, 'production' refers to processes of molecular interaction that generate new molecules which then participate in further interactions. It is clear from the discussion of structuration that actors do not produce structure anew but rather reproduce or transform that which already exists. However, I do not see that this is incompatible with a notion of production since one could say that molecular production does not create something from nothing, but simply reorganizes or recombines components (atoms and molecules) that already exist. One significant difference from the Nomic example is that in the game the moves are intended to change the rules – that is their primary purpose – and the players will be conscious of this, whereas most if not all social activity is *not* intended to reproduce structure; this is merely an unintended and probably unrecognized consequence. Again, I do not see that this invalidates the notion of production – all that is necessary, and indeed both Giddens and Maturana stress this non-functionalist view, is that (re)production of structure actually occurs. If it does, whether intended or not, autopoiesis continues; if it doesn't, the particular social practices will die out.[34] A tentative conclusion thus far is that we can, contra Varela, identify components and processes of production.

The second major requirement of autopoiesis is that the system is organizationally closed and generates its own boundary.[35] This means that the processes in the network must feed back upon themselves to form a circular concatenation and thereby implicitly demarcate the system from its surroundings. In the case of physical autopoiesis the boundary would be spatial and would involve specific components (e.g. the cell wall), but – as Varela points out – this is not necessary in the more general case of organizational closure, where the nature of the boundary will depend on the type of components involved.

Whether this condition is satisfied is harder to answer in the case of social systems. Taking first the question of closure under some type of circular relationship, there is clearly a form of closure between the social structure in general and the social activity through which it is (re)produced. However, this is rather different from the circularity of physical autopoiesis, where molecules interact with other molecules to produce yet more molecules. In the social case, the relation is between two different strata – social structure and social action – rather than within the one stratum.[36] To be strictly analogous to the physical example we would have to look for circularity among the elements of structure – position-practices and rules producing more of the same. No doubt there are many relationships between these components, indeed Bhaskar (1979: 41) defines them relationally, but since social structure only exists through social activity, positions and rules cannot simply produce themselves. This situation is clear in Nomic – the rules are only transformed through the activities of the players. This of itself does not preclude organizational closure, but we would have to accept that social systems are different from material systems in the ways Bhaskar suggested above.

Apart from the general notion that action (re)produces structure, we can also see many specific circular feedback loops involved in this process. Giddens distinguished three different types – homeostatic loops via unintended consequences of action, self-regulation through information filtering, and reflexive self-regulation involving conscious manipulation of social institutions; Giddens uses the poverty cycle as an illustration of all three. We could obviously look empirically at any part of society and discover an enormously complex intermeshing of causal loops involving both observable activity and events, stretching over time and space, and the underlying structure of positions and rules. The difficult question, though, is to what extent such circuits can be said to form a boundary, or at least demarcate themselves from the background. This is a strong but necessary feature of organizational closure as defined, since it is what accounts for the systems' identity and its domain of possible interactions as a whole. In an interesting paper about social boundary mechanisms, Varela argues:

> Thus a unity's boundaries, in whichever space the processes exist, is indissolubly linked to the operation of the system. If the organizational closure is disrupted, the unity disappears. This is characteristic of autonomous systems.... It is also apparent that once a unity is established through closure it will specify a domain with which it can interact without loss of identity.
>
> (Varela 1981a: 15)

We can see how this applies to physical systems such as the nervous system or the immune system (Varela *et al.* 1988). In the case of a non-physical system, if it is well defined like Nomic we can say that at any point the system is able to distinguish inside from outside – valid rules from invalid ones. But it is not obvious that we can actually identify such clear-cut examples as Nomic within the mêlée of society as a whole. There are many different possibilities (Giddens 1990b, p. 303): nations, states, or perhaps societies as such; Western capitalism as a whole; enduring institutions such as religions or political parties; particular collectivities such as firms, clubs or social movements; small-scale groupings such as a family or a sports team; or, following Luhmann, functional subsystems such as the economy, law, and politics. Considering what might be the boundaries of a society and what could be its domain of actions as a unity can indicate the difficulties. Tilly (2004) examines a number of different potential mechanisms for generating and maintaining some form of separation or boundary between different social groups.

Giddens (1981: 45) has suggested three criteria for a social system to be considered a society: (i) an association with a particular time-space location with a legitimate claim to make use of it; (ii) a shared set of practices involving both system and structure; and (iii) an awareness of a shared identity. In terms of time and space, societies will be localized to some extent and, especially in historical times, there may well have been particular examples such as nomadic peoples or forest tribes who were genuinely self-contained. We can look back and see different societies clashing with each other as in periods of colonization. But in the modern world, with its tremendous global interpenetration through communications and transport, is it possible to draw any such lines any more? Societies certainly don't coincide with nation-states, being both wider, e.g. European society, and narrower, e.g. Scottish and English. Indeed it can be argued (Angel 1997) that nation-states themselves will become of lesser importance than global companies. Luhmann (1982b) concluded that one had to go up to the level of the world society as a whole.

We can also to some extent pick out enduring social practices, but at whichever level we look at these are many and diverse. There may be greater differences within a notional society than between that society and another, especially with the tremendous intermixing of ethnic and cultural groups within modern societies. A sense of identity may be equally polysemous – one could feel Mancunian, English, British, European, or Western depending on who one was interacting with. As Giddens concludes,

> It is important to re-emphasise that the term 'social system' should not be understood to designate only clusters of social relations whose boundaries are clearly set off from others. … I take it to be one of the main features of structuration theory that the extension and 'closure' of societies across space and time is regarded as problematic.
>
> (Giddens 1984: 165)

6.4.4 Summary

This section has developed the following conclusions concerning the application of autopoiesis to social systems as seen from a broadly structurationist perspective.

In terms of components and processes of production,

- we can take the components of such a system to be those of social structure developed above – rules, resources, positions and practices;
- we can identify processes of production (in terms of reproduction and transformation) of these components provided we accept that with social systems production involves the transformation of an existing structure, and a duality between social structure and human activity.

In terms of organizational closure,

- we can identify a circularity of relations both in the generic (re)production of structure and in specific causal chains;
- but it is difficult in general to identify specific social systems that are clearly bounded and have identity. This may be possible in specific, well-defined instances (for example, Nomic), but this would require empirical verification.

Thus we cannot conclude in general that social systems, conceptualized as a synthesis of structuration and critical realism, *are* autopoietic. Nor can we follow Varela and say they are not autopoietic but organizationally closed. However, most of the key elements of self-producing systems can be seen in social systems, and it may be that particular examples could embody them all.

6.5 Conclusions

The purpose of this chapter was to evaluate in detail the extent to which social systems could be conceptualized as self-producing, autopoietic, in an ontological rather than simply metaphorical sense. The first step was to specify clearly what we took to be the essential core of the theory of autopoiesis – a specification of particular components that participate in processes of production of similar components within a well-bounded whole. The next step was to consider the extent to which autopoiesis was compatible with, or contributed to, existing social theories. For this purpose Luhmann's communication theory, Giddens' structuration theory, and Bhaskar's transformational model of social activity have been examined.

In the case of Luhmann, the conclusions were that his social theory did consistently embody a version of autopoiesis although this version is not wholly compatible with Maturana's original formulation. The components were clear (communications) and a mechanism was specified for generating closure, but the production processes and the supposed isolation of various systems was considered problematic. However, this theoretical purity was obtained only at the expense of a very abstract and impoverished view of social processes and interactions.

In the case of structuration theory, we had first to construct a synthesis from two different versions developed by Giddens and Bhaskar respectively. The conclusion then was that components and processes of production could be identified (rules, resources, positions and practices), but that it was extremely difficult to identify empirically the bounded closure of a particular social system.

Thus, the overall conclusion is one of agnosticism. Autopoiesis as a social theory has many attractions, and there may be very specific social situations, exemplified by Nomic, where it could be identified. But, in general, I do not believe that social autopoiesis has yet been demonstrated. Nevertheless, further research in this area is certainly to be encouraged, which might take various forms:

- Attempting to demonstrate empirically a self-constructing social system along the lines described in the paper.
- Developing further Maturana's other theoretical ideas concerning the biological basis of observation, languaging, and embodied cognition. This leads to a particular view of interacting human agents at the individual level which could possibly be combined with modern complexity theory (Byrne 1998) at the system level to produce an interesting new synthesis.
- Synthesizing Giddens' and Luhmann's theoretical systems, which seem to me to be potentially complementary. They could be developed as an orthogonal pair of distinctions – that between observable system and underlying structure (Giddens) on the one hand, and, within the system, that between individual interaction and societal communication (Luhmann) on the other.

Whether or not social systems can be seen as autopoietic, there are several other approaches that look at social systems from a mechanistic viewpoint, including Mayntz (2004), Wright (2004) and Sawyer (2004).

Notes

1 Hayles (1999) provides an insightful discussion of the transition from the positivism of first-order cybernetics to the constructivism of second-order largely generated by Maturana and Varela's theory of the observer. Wolfe (1995) presents an interesting discussion of the implications of this transition for feminism and humanism.

2 There are other aspects of the general theory of autopoiesis that are relevant to social theory, for instance Maturana's theories of cognition, language, self-consciousness, consensual domains, and domains of explanation, but these are not addressed in this chapter.

3 There are two major dimensions to the debate. First is that between agency and structure as to the ontological validity of concepts of system or structure at all. The anti-structural position has been put forcefully by King (1999a, 1999b, 2000) and rebutted by Archer (Archer 2000b). Second are debates between various proponents of different structural approaches. Of particular relevance here are debates between Luhmann and Habermas (Habermas 1985) and between advocates of Giddens' structuration theory (Cohen 1989; Manicas 1998; New 1994) and Bhaskar's critical realist social theory (Archer 1996; Archer 1998; Bhaskar 1979; Porpora 1989).

4 The basic theory for living systems and cognition is expounded in Maturana and Varela (1980), with a version written for a wider audience in Maturana and Varela (1987). One of the most recent comprehensive expositions is that of Maturana *et al.*

(1995). Maturana has also written on the implications of his ideas for social theory (Maturana 1980b), science (Maturana 1990), aesthetic experience (Maturana 1993), and self-consciousness (Maturana 1995). Mingers (1995b, 1997a) provides a critical introduction to autopoiesis and its application in different disciplines. There are special issues of *Family Therapy Networker* (9:3, 1985), *Irish Journal of Psychology* (9:1, 1988), *Cardozo Law Review* (13:5, 1992), *Cultural Critique* (Spring 1995), *Social Science Information* (35:2, 1996), and *Acta Sociologica* (43, 2000) devoted to autopoiesis and especially the work of Luhmann. Much of his theory tends to be rather abstract, but Vanderstraeten (2000) looks at the implications for socialization theory and Paterson and Teubner (1998) try to demonstrate legal autopoiesis empirically.

5 In the beginning, in fact, the open systems concept was developed in contrast to the prevailing classical tradition, especially in physics, of *closed* systems that did not interact with their environment at all (von Bertalanffy 1971). This should not be confused with autopoietic systems, which are *organizationally* closed but *interactively* open (see later discussion of this distinction).

6 Although it is by no means the only way of conceptualizing organizations – see Burrell and Morgan (1979); Morgan (1986).

7 This is very much the basis of Luhmann's functionally differentiated society to be discussed later.

8 For the moment we shall assume physical autopoiesis. An example to have in mind might be a single-celled organism such as an amoeba.

9 Commentators often talk of *re*production, but there is an important distinction to be made. In the physical domain *re*production implies the creation of another, separate, entity. This is *not* what autopoiesis means, for it is about the continual self-production of the same entity. In the social domain, Giddens and Bhaskar talk of the 'reproduction of society' meaning by this the continuation (or transformation) of the same society in a similar way to Maturana. In this chapter I shall, where appropriate, write (re)production to mark the distinction.

10 Historically, Luhmann has been enormously prolific over 30 years, although much remains untranslated, and he is best known outside Germany for his debates with Habermas (e.g. Habermas 1985; Habermas and Luhmann 1971). Much of his work is developed from, and a reaction to, Parsons in two ways: explicitly bringing in phenomenology through the category of 'meaning'; and at the same time radicalizing the abstract and autonomous nature of 'system'. A major step in his development was to reformulate his theory from an action to a communication orientation (Stichweh 2000), followed by a further reformulation into autopoietic terms. Several papers outline his use of autopoiesis rather schematically (Luhmann 1982b, 1985a, 1985b, 1986, 1987a, 1992).

11 His theory has been applied in other areas, especially law and public administration (Brans and Rossbach 1997; Luhmann 1987b, 1989b; Teubner 1993; Teubner and Febrajo 1992; Veld *et al.* 1991).

12 The particular nature of 'society' for Luhmann will be brought out later.

13 Indeed, he argued this before any mention of autopoiesis; see Luhmann (1982a: xx).

14 Communications, Luhmann argues (1986: p. 177), are more fundamental than individual communicative acts. This is because first, actions need not be inherently social whereas communications are; second, social actions already presuppose communications in the sense that they rely on or raise the expectation of recognition, understanding and acceptance by others. In other words, a social action is inevitably already a communication. Yet, third, a communication is more than simply an action. It involves and therefore includes the understanding of another party and so goes beyond the individual action to form the link necessary for social operations. A communicative act in itself leads to nothing; it is only when it generates some understanding in another that it can trigger a further communication.

15 A major distinction made by Luhmann is between *society* and *interaction*. Interaction is direct and face-to-face between people. Society emerges from the undifferentiated domain of interaction and provides the capacity for closure – it *structures* communication (Luhmann 1995: 416).

16 For example, a discovery in science such as genetic manipulation may lead to political decisions about limits of acceptability resulting in new laws governing its use and eventually economic activity, each of which feeds back to the scientific subsystem. Moreover, these effects may be disproportionate between subsystems – a minor occurrence in one may trigger a major response in another.

17 See also Teubner (1993, Chapter 2).

18 Giddens himself mentions autopoiesis:

> The most relevant sources of connection between biological and social theory … concern recursive or self-reproducing systems. There are two related types of theory involved here … [the theory of automata] … is not of as much interest to the conceptualisation of social reproduction as recent conceptions of cellular self-reproduction (autopoiesis).
>
> (Giddens 1979: 75)

19 Giddens (1981: 30). To unpack this slightly, these three relations/differences are the *where*, the *when*, and the *what*. Space and time are straightforward and can be seen as syntagmatic dimensions. The third dimension of difference is paradigmatic – that is, a specification or selection from a domain of differences. For Giddens, the first two relate to system and the third to structure.

20 Very briefly, my argument against the hermeneutic critique is the obvious one that understanding is never transparent to itself in terms of either its grounds or its consequences. With regard to causality, we can accept that people are the only source of intentional, efficient causation but, as with any system (including physical ones such as amoebae), the parts act in a way that generate the emergent properties of the whole but, at the same time, the configuration of the whole shapes the behaviour of the parts. With a more Aristotelian view of causation we can accept that only people act (efficient cause), but that society shapes that action (material and formal cause). This is the view that was articulated in Chapter 5.

21 In this, I largely agree with New (1994) but disagree with Wright (1999), who takes criticisms of Giddens largely on trust.

22 Bhaskar contrasts this with a 'dialectical' view of the relationship (as advocated by Berger and Luckmann (1967) which, he claims, sees people and society as two moments or sides of the same process, rather than as two distinct, but interacting, objects. This is a criticism that has been applied to structuration theory (Archer 1996).

23 Bhaskar relates his work to Giddens (Bhaskar 1979, p. 35) and Giddens also uses Bhaskar's arguments (Giddens 1984, p. 340).

24 This is repeated in one of Bhaskar's later discussions of the social (Bhaskar 1994, p. 95.)

25 Although Giddens might disagree on the extent to which 'society' can be clearly identified – see below.

26 Interpreted in Giddens' general sense of procedures for enacting practices (Giddens 1984: 21).

27 Cohen (1989: 209), one of the main interpreters of Giddens' work, also suggests that Bhaskar's notion of position-practices could usefully be incorporated in Giddens' structure.

28 An interesting anomaly in Giddens' concept of structure has been pointed out by Sewell (1992). It is said to consist of rules and resources, and resources can be authoritative (power over people) or allocative (power over objects). Allocative resources are themselves material, e.g. raw materials, technology, goods (Giddens 1984: 258), and so how can they be part of structure which is virtual?

29 For an analysis of the effects of change and reflexivity in late modernity, see Giddens (1990a; 1992).

30 In a note at this point Giddens refers approvingly to Bhaskar.

31 And sound very much like Bhaskar's generative mechanisms.

32 Giddens (1979: 78). These are taken unchanged from systems theory, where they would be called multiple cause feedback loops.

33 Nomic is a game developed by Suber (1990) to illustrate the self-referential nature of law – 'only laws can make laws'. In it, players take turns where a turn consists of proposing and voting on changes to the rules of the game itself.

34 This aspect of social reproduction, although not stressed by Giddens, is easily observable. As technology develops, old practices die out simply because they no longer occur and are therefore not reproduced.

35 To clarify the difference between organizational closure and autopoiesis per se, the latter is simply a special case of the former. Organizational closure occurs when processes within a system become circularly linked to each other, thus generating an entity that has a degree of autonomy in defining its own boundary. These processes can be any, e.g. descriptions, computations or productions (Mingers 1997b). When the process is one of production, the systems is autopoietic (Varela 1981a). The definition of organizational closure is identical to that given for autopoiesis but with 'interaction' substituted for 'production'.

36 Mathematically, closure can be clearly defined. A particular domain of objects is closed with respect to a particular operation if the result of the operation always remains in the same domain. Thus the domain of positive numbers is closed with respect to addition but not with respect to subtraction.

Part III

Epistemological issues

7 Observing systems

The question of boundaries

7.1 Introduction

In the last three chapters we considered ontological issues around the concepts of mechanisms and systems. In Chapter 4 we developed the view that the world consisted essentially of mechanisms or systems that were spatially and temporally located, and that had powers and tendencies to behave in particular ways. The interaction of these systems generates the occurrences in the world. In Chapter 5 we looked in more detail at what was meant by powers and tendencies, and how this form of generative causality might actually work. In Chapter 6 we considered the special characteristics of social systems that distinguish them from physical systems. In particular, we examined the idea that social systems might be a form of self-producing or self-constructing, autopoietic system in a strong ontological sense.

In this chapter and the next we are going to move from the ontological to the epistemological and consider issues around the question of how we might come to have knowledge of such systems, and what knowledge itself might be. This issue is indeed at the heart of epistemology – what can we say about the relationship between our putative knowledge of the world, as represented by our mental perceptions and conceptions, and the external world itself? There are many different views here, ranging from naive realism, which sees the relation as one of unproblematic correspondence, through to extreme forms of constructivism and phenomenology that argue that we actually have no access to the external world at all and live in a solipsistic world of our own construction. This is very much the domain of critical realism, which maintains a non-naive form of realism. It has also been a major preoccupation of systems thinking with the epistemological break between first-order and second-order cybernetics or, equivantly, hard and soft systems thinking.

We will approach this by considering in some detail the notion of a system boundary. It is an interesting paradox that a system boundary is one of the most fundamental concepts underlying systems thinking and yet it is one of the least discussed, especially in the seminal literature. Arguably, the concept of a 'system' existing within an 'environment' is the foundation for systems theory. What is it that separates a system from its environment? – the system boundary. In fact, defining a system in terms of its components and their relations is effectively to delineate its boundary. Or, put the other way, in order to define a system it is necessary to define its boundary. Thus the drawing of a boundary is in fact the most primitive systemic act that one can perform. A boundary is also, arguably, of great importance for the mechanistic view as discussed in Section 4.3, for generally causal mechanisms have to be identifiable and locatable within some kind of space.

In this chapter, I want to explore this question in an almost phenomenological way by considering what we can learn of the nature of boundaries in a whole range of different types of systems. Initially, I shall 'bracket' the question of the observer and simply try to discover what properties boundaries have across a range of different systems. I shall begin with what seems the simplest – the boundaries of easily identified medium-scale physical systems. I shall move from physical boundaries to conceptual boundaries, beginning with mathematics, where one might expect there to be the greatest concern with definition and precision, and then going to more general conceptual boundaries, particularly linguistic. From there I shall consider the difficult question of boundaries for social systems. At this point I shall deal directly with the question as to whether boundaries are in fact purely observer constructs or whether we can in some circumstances take them as having independent ontological existence. As might be expected, given the critical realist stance, I shall come down on the side of existence. The chapter will finish with a consideration of systems which may bound themselves through organizational closure, and finally the setting of boundaries within organizational research and intervention.

7.2 Physical boundaries

To try to understand the nature of boundaries I shall begin with physical objects, which would seem to be the most straightforward cases to deal with, moving on to conceptual, social and mixed cases.

7.2.1 Basic forms of boundary

Type I – edges and surfaces

The first and most primitive form of boundary is what we might call an edge or surface. That is, it is simply the limit of the extent of some entity or substance. Examples which have such boundaries might be a pool of water, a table, a shadow, or a sheet of paper. In each case the edge marks the transition from or difference between one substance and another (Bateson 1973a). With an edge there are no specific boundary components, just a transition from one thing to another. To be more precise, we need to distinguish an edge from a surface.[1] A surface is a two-dimensional area of some extent. If it joins another surface then it forms an edge (and if two edges join they form a corner point). In two dimensions a surface is bounded by its edges, although some surfaces do not have any edges, e.g. a sphere, and are therefore unbounded (but finite!). In three dimensions a volume is bounded by its surfaces, e.g. a cube. A surface, such as a sheet of paper, can be seen to have two sides as well as its edges. These also form part of the boundary of the object as a whole, since they are the interface between the object and another substance, e.g. air or a table.

It is the case that edges (and surfaces) are always fuzzy to some degree. Although an edge may appear to the observer as being very sharp, if looked at a high enough resolution it will be seen to be imperfect. Some edges are, in any case, clearly imprecise either because of their rough (in a fractal sense) nature, e.g. rocks; or because of their dynamic behaviour, e.g. water lapping on the shore.

These latter observations show that what can be detected as an edge (or any other type of boundary) is always relative to time and to the level of resolution. In general (and this is true of the systemic differentiation of structure and process), as we lengthen the period of consideration (at a given level of resolution) more will become dynamic and the boundary will become more fuzzy. The more we shorten the period, the more things will be unchanging and the more definite the boundary will be. Consider waves lapping on the sea shore. If we take a snapshot we will see a relatively fixed boundary between the water and the sand. As we lengthen the time period from seconds to minutes to hours the boundary becomes more variable, especially as the tides change. The same is true of resolution level. Seen from a distance the edge of the paper is sharp, but as we magnify it more and more to an atomic level the original boundary virtually disappears.

Type II – enclosures

An edge simply demarcates or separates one thing from another. An enclosure or container is different in that it consists of specific boundary components that both mark a separation and keep in that which is included; or, equivalently, keep out that which is excluded. Examples are a bottle, a bubble, a circle drawn on paper, an onion skin, a football, a suit of armour, insulation round a wire or pipe, or a fence. The important characteristics are:

- The enclosure consists of specific components different from, although related to, both inside and outside components. An enclosure has two surfaces, one facing in and one out.
- The enclosure is *relatively* impermeable, but never completely. It prevents the movement of certain substances but not others. A bottle contains liquids but not necessarily gases or radiation; a fence contains sheep but not insects or birds. The enclosure does not have to be complete to have an effect – an open bottle is still a bottle – although the effect may be different, e.g. the liquid may evaporate.

Type IIa – membranes

A membrane is a special example of an enclosure. It is an enclosure that is active rather than passive. In particular, it is a biological enclosure, which is one that is part of a living organism. Examples are the cell wall, a person's skin, or an artery. A membrane is distinct from an enclosure in being part of the autopoietic structure, in that its components are produced and maintained by the organism in an on-going process. It is thus self-constructed by the autopoietic system rather than simply occurring as a result of physico-chemical processes as do a bubble or a gravy skin.

Type III – demarcations

We also need to consider the question of boundaries for physical systems that do not occupy contiguous areas of space. To what extent can they be said to have a boundary? Consider, for example, the solar system, a pendulum in a magnetic field, or a central heating system. The solar system is a system by virtue of the gravitational effect of the sun but it has no boundary components. Pluto may be the most extreme planet but there could potentially be others (indeed a new planet was claimed in 2004) as the gravitational force extends further out although decreasing in strength quadratically. So the solar system does have a boundary surface or edge but this is constituted in terms of the limits of a force rather than a substance.

Consider a pendulum system that consists of a base with battery-driven magnet and a support holding a pendulum with a magnet over the base. The magnets repel each other and the pendulum undergoes chaotic (i.e. random) motion. Assuming that it is the dynamic behaviour that we are interested in, then clearly the system needs to include both the physical components and the magnetic force connecting them. This means that there is not a contiguous edge or boundary physically surrounding the system. Instead the boundary becomes more of a notional device for separating conceptually what belongs to the system and what does not.

Considering next a central heating system, there are potentially many different aspects (or systems) we could be interested in – the system that maintains a constant temperature in the house; the system of flows of water through the pipes; the system consisting of the boiler and its gas supply, and so on. Each will consist of a different set of elements, not all of which will be physical objects. For instance, the heat regulation system consists of the boiler, pipes, water, radiators, and thermostat, but in order to work as a regulation system it needs to engage in information processing – comparing the information about actual temperature with desired temperature and then sending electronic signals to the boiler. Again, the various different boundaries pick out, for different purposes, the members of different systems.

In these three examples we can see that systemic thinking involves more than the simple recognition of individual objects. It begins with a particular phenomenon to be explained or purpose to be achieved. It then requires a degree of conceptualization, rather than mere perception, to characterize an appropriate system in terms of components, relations and boundary. The boundary may in part have a material embodiment, but generally it will simply represent a distinction or demarcation between that which has been selected as part of the system and that which has not. This does not mean that the boundary is purely arbitrary, or is wholly a construction of the observer. It rests on the components and relations that exist independently in the intransitive domain even though it is selected by the observer. This is demonstrated by the fact, already mentioned in Section 3.3, that the observer may make the wrong selections. Knowledge is fallible, and we will find out from the real world whether our choices of components, relations and boundaries yield the appropriate behaviour.

7.2.2 Multiple boundaries

So far we have assumed that an object has just one boundary, but that is in general not the case. Systems may have many boundaries depending on the nature of the interactions they are undergoing. For instance, if we consider the body it is usually seen as bounded by the skin, and this is so for many interactions, for example with light or with touch. But the skin is not a boundary for X-rays or for a very sharp knife, for which the outside of the bones form a boundary. Equally, the skin does not form a boundary for particular chemicals, or for very small organisms that can enter the pores.

To be accurate, therefore, in specifying a boundary we should also specify classes of interactions or agents for which the boundary is a boundary. In everyday language this is often assumed, or obvious from the context. It becomes more of an issue with complex social systems, such as a university, which can have many boundaries in different dimensions.

An alternative to specifying a boundary in terms of particular interactions is to specify it in terms of the purposes of the observer. In general, an observer will have chosen some aspect of their experience which they wish to explain or perhaps change, and this will shape the nature of the system or boundary that they focus on. This point will be discussed more extensively below.

7.2.3 Natural wholes

The previous section raizes a deep question. Are there some systems that form natural or integral wholes (Simons 1987) which observers may *discover* and then others, less tightly connected, where it is the observer who delineates the system?

For Maturana and Varela (Maturana 1970a; Maturana and Varela 1975, 1980), it is fundamental that we can treat a system either as a unity, that is an unanalysed whole, or we can study its parts and their interactions. These are distinct and irreducible domains. This certainly suggests that for them systems are natural wholes. Concepts such as structural coupling and structural determinism also presume that a system can be well-defined and separated from its environment. This seems very natural for biologists, since organisms would obviously be prime examples of self-constructed wholes (see the extensive discussion of wholes and emergence in Section 5.2.1).

I propose to take their basic argument but apply it to the intransitive domain. We can then argue that natural wholes exist in so far as the following requirements are met:

- The system possesses characteristic(s) or behaviour(s) that are only attributable to the system as a whole by virtue of it being a whole. They are not attributable to the parts. This implies that the system must be able to be clearly distinguished as a whole separable from its environment.
- The system consists of parts and relations between the parts (its structure) that together are necessary in order to generate the characteristics of the system as a whole. In some cases this will include specific boundary components, but this is not a necessity.
- The relationship between parts and whole is recursive – the parts may themselves be whole systems.

Note that this does not imply that parts may be a part of only one system – they may be parts of several systems simultaneously. For instance, blood vessels are part of the blood circulation system and the immune system; my computer is part of my office system but also part of the university computer network. Equally, different systems may occur across the same set of parts depending on what particular characteristics are of interest.

Clearly there are many systems, both natural and designed, that meet these criteria and that would generally be recognized as wholes even if they do not have specific boundaries. But as we move from the physical domain to the social and economic world there is a much more complex level of interaction between many different kinds of entities, and wholes do not separate themselves out so clearly. Systems identification is much more dependent on the choices of an observer, as will be discussed in Section 7.7.

We can summarize this section on physical boundaries by noting:

1 Some characteristics of boundaries:
   - *a* What we observe as a boundary is always relative to the space and time frame of the observations. As these change, different boundaries come to be presenced. But this is an epistemological point to do with our observations – the boundaries still exist ontologically.
   - *b* Boundaries are inherently fuzzy if we take a high enough degree of resolution.
   - *c* Equally, boundaries are never perfectly impermeable – they will always be open to some elements and closed to others.
   - *d* Any particular physical object(s) will have several potential boundaries, depending on the nature of the interactions or the purposes of the observer.
   - *e* Moving away from single physical objects, the system will contain a variety of elements depending on the observer's purpose and the boundary will become more notional than actual.

2 The effects of boundaries:
   a Separation or demarcation of different substances, elements or spaces.
   b Containment or inclusion/exclusion of substances, elements or spaces.
   c Self-production and functionality, e.g. permeability.

7.3 Mathematical boundaries

If we move away from physical systems to conceptual ones, then the most precise and well-defined area of thought is probably mathematics. Here we can find many examples of boundaries, especially in terms of 2*a* and 2*b* above, i.e. separation and containment. Not being a mathematician, I shall treat these informally.

7.3.1 Mathematics of shape

Geometry

The most obvious area of mathematics relevant to boundaries is that of geometry, which is the study of the properties of forms or shapes within Euclidian space,[2] for example a circle on a plane surface. One of these properties is that the shape will usually divide the space into separable regions – inside and outside – and will thus form a boundary. Mathematically, these boundaries are perfect in the sense that they are infinitely thin and we can unambiguously determine whether a point lies inside or outside. Indeed, they can be defined algebraically rather than pictorially. The equation of the circle ($x^2+y^2=a^2$) is effectively a set of instructions to determine whether a particular point is inside, on, or outside the circle.

Geometry is obviously a good model for the sort of physical boundaries we have examined above, although when it is applied in the real world, e.g. drawing circles or measuring distances, degrees of fuzziness come in. In fact traditional geometry is rather strict in many of its assumptions and it is interesting to consider two developments that relax some fundamental ones – topology and fractal geometry.

Topology

Topology is a kind of non-metric geometry (Flegg 1974) in that it also studies the properties of shapes but without considering distance or measurement. So in topology the circle, square, triangle or indeed any random shape that forms a closed curve are all considered the same. One property that remains central in topology is that of separating a plane into distinct regions, i.e. drawing a boundary, although surprising results can occur. For instance, if you draw a circle on the surface of a sphere it separates the surface into two regions and you cannot move from one to the other without crossing the line. The same is not always true on the surface of a torus, that, is a 3-D ring. If you draw the line around the torus, as if you were going to slice through it, then it does not create two regions since any point can be joined to any other without crossing the line. Thus some closed boundaries do not actually enclose anything.

Topology also investigates the properties of the surfaces the shapes are inscribed on. Usually a surface, such as a piece of paper, has two sides separated by an edge (a boundary curve) that must be crossed to move from one side to the other. But it is possible to construct interesting surfaces such as the Mobius strip, which has an edge but only one side, and the Klein bottle, which has only one side and no edges (an ordinary bottle has an inside and outside, and one edge – the rim).[3]

Bunge (1992), in one of the few recent contributions that aims to define the concept of system boundary, based his characterization on topology, in particular, using the concepts of topological space and neighbourhoods. A topological space is a very abstract concept concerning the relations between the members of a set of elements of any kind. Given a set (for example, of components of a system) the elements can be arranged in different groups or subsets. When taken together, open subsets do not include their boundary elements, while closed subsets do. For instance, the set of integers between 0 and 10 exclusive (i.e. 1–9) is open but the integers >= 1 is closed since it includes its boundary element (1). A neighbourhood of a particular point, *p*, is a subset of elements that contains *p* and that is contained within an open set. Elements can have several neighbourhoods. From concepts like this we can generate the interior of a set (excluding its boundary), the closure of a set (the interior plus boundary), the exterior of a set (the interior of the complement of a set) and the boundary (that which is neither interior nor exterior).

From these ideas, Bunge identifies the set of interest (relative to a given system) as consisting of all the components of the system and its environment. Relations between components are based on the idea of one component acting on another; and a neighbourhood of *p* is the set of components directly linked to (i.e. related to) *p*. The boundary components are then defined as those components which have neighbourhoods including both system and environment components. Elements whose neighbourhoods only contain other system components are part of the system, and elements whose neighbourhood only contains non-system components are in the environment.

How useful is this type of definition? First, several technical flaws have been identified by Marquis (1996), who provides a revised version based on the same general principles. However, from our point of view, that of identifying the boundaries of systems in the real world, this approach provides no help whatsoever. It begins with the assumption that there is a known set of system and environmental components and proceeds to deduce the boundary from these. But the whole point is to be able to identify the system in the first place by drawing its boundary. The identification of the system and the drawing of a boundary are two sides of the same act. If one has been done the other simply follows on. Indeed, even Marquis recognizes this, concluding that:

> It is in fact tempting to say that the real problem has to do with the criterion for the boundary, that is, how we actually find in actual particular cases the boundary of a given system and that this question has no general solution. It is simply not the task of general systems theory to answer such a question.
>
> (Marquis 1996, p. 254)

But that is indeed the question that we wish to address!

Fractal geometry

The second development is fractal geometry, which allows lines to be rough rather than smooth as is conventionally assumed (Mandelbrot 1982). Consider a coastline. When viewed on a map it is clearly not a smooth line but jagged and broken. You might think that if you looked at it at a different scale, say from the air, or down on the ground, it would become smoother. But this is not in fact the case. Remarkably, it has the same degree of roughness at whatever scale you observe it. Such lines are said to be self-similar at all scales. Real-world shapes such as coastlines, mountains or rivers are statistically self-similar in that they are essentially random rather than ordered. Lines can display different degrees of roughness. A straight line between two points has a fractal dimension of 1. As the roughness increases the fractal number increases towards 2, the dimension of a surface or area. The Norwegian coastline has a fractal number of about 1.6. Shapes in 3-D such as mountains typically have dimensions of about 2.5, and a cube would be 3. Within fractal geometry, therefore, boundaries can be indefinitely fuzzy.

7.3.2 Sets and operations

It has been argued since *Principia Mathematica* (Whitehead and Russell 1925) that sets may form the basis of mathematics, and what is a set but a separation of a group of elements from everything else. A set is simply a list of elements, in no particular order, defined by a criterion of membership which is either *intensive* or *extensive*. An intensive criterion defines membership by a specific meaning or connotation, as in 'the vowels in English', whereas an extensive definition names or denotes individually all the members, as in {a e i o u}.

The extensive definition is clearly the more precise, although even here it would be possible to make a mistake and wrongly include or exclude something. The intensive definition is the more general and useful, but this ultimately rests on language; no matter how tightly definitions are drawn there will always be a degree of equivocation even in mathematics (e.g. Gödel's theorem), let alone in ordinary language.

What is different from the point of view of boundaries is that there is no actual boundary as such at all. No members of the set are marked out as boundary members; none can even be said to be closer to the edge in some sense, nor is there any visual representation as there is in geometry or topology.

All we can say is that a distinction has been drawn (Spencer-Brown 1972) and that it allows us to separate elements into members and non-members.

Having defined a set we can also consider operators or functions that can operate on members of the set. In a system's sense this is akin to moving from a static or structural view to a dynamic or processual view. The question is, when we apply the operator to one or more members of the set does the result also belong to the set or does it cross the notional boundary and become a non-member? In other words, is the set closed under that particular operation? To give an example, the (infinite) set of positive integers is closed under addition but open under subtraction. Adding two positive integers always gives another positive, but subtracting sometimes gives a negative which is not then a member of the original set.

Traditional set theory assumes that the boundary of the set is exact, i.e. that any element can be unambiguously assigned as a member or not. In practice this is often not the case, and this has led to the development of fuzzy set theory (Zadeh 1965). Suppose we are interested in the set of 'tall' people. We could define tall arbitrarily as over two metres. This gives a hard boundary but does not really capture normal usage. Fuzzy set theory would instead define a point that was definitely not tall and a point that was definitely tall, and then have a membership function between the two points ranging from zero to 100 per cent.

We can summarize this section on mathematics by noting the following:

- Geometry provides a formalism for considering boundaries, but it is too idealistic for dealing with real-world boundaries. Developments of geometry do go some way in this direction – topology only considers the relationships between shapes and not their distances or sizes, while fractals allow boundaries to be rough rather than smooth. Both still assume that the boundaries (lines) are perfectly impermeable (or 'continent', to use a term from Spencer-Brown to be discussed in the next section).
- Set theory introduces the idea that elements can be separated without an actual boundary at all by a membership criterion. Moreover, operators can lead to entry or exit, i.e. input or output; a crossing or not crossing of the notional boundary. As in geometry, set theory assumes a perfect distinguished boundary, but this is relaxed somewhat by fuzzy set theory.

7.4 Conceptual boundaries and language

Given that a boundary is primarily about separation and difference, then we can see that language, from individual concepts right up to whole languages, is essentially a play on boundaries. This is the insight of the semiotic linguists from Saussure (1960, originally 1916) to Derrida (1978), not to mention cyberneticians such as Bateson (1973a). Perhaps the most interesting starting point, and one which provides a link between mathematics and language, is that of George Spencer-Brown's (1972) *Laws of Form*.

7.4.1 Spencer-Brown's laws of form

Spencer-Brown's primary aim was to uncover what lay underneath logic and, in particular, Boolean algebra.[4] The latter was created to provide a mathematical analysis of binary logic. Logic can be defined as the science of abstract form (Lee 1961, p. 13), where form is taken to be organization, pattern, structure, or relationship. Form, as such, cannot be shown separate from a particular content but it can be studied, in abstraction, through an appropriate symbolism. Thus logic can be seen as the study of relationship or order in general, abstracted from any particular content. Spencer-Brown's work was aimed at 'the form in which our way of talking about our ordinary living experience can be seen to be cradled. It is the laws of this form, rather than those of logic, that I have attempted to record.' (Spencer-Brown 1972, p. xxiv.)[5]

His approach began with the realization that Boolean algebra, as the name implies, is purely an algebra and that no one has ever studied the arithmetic upon which this algebra is based.[6] He therefore set about trying to discover the primary, non-numerical arithmetic for Boolean algebra. Logic and Boolean algebra concern the form of linguistic statements. To find an arithmetic therefore means going beneath the level of language to uncover that upon which language itself rests.

Spencer-Brown, like Maturana, sees language as essentially practical, not purely descriptive, and takes the most primitive activity as that of *indication* or *distinction*. To distinguish something is the most basic act we can perform. Before counting things we must be able to distinguish between them, and before distinguishing several different things we must be able to distinguish something. This is the basis of all language – to create from nothing ('the void') one thing, or state, or space which is distinct. The laws of form are concerned with the consequences of this most primitive act – the act of drawing a distinction. Spencer-Brown prefaces his book with a quotation from Lao Tzu: 'The nameless is the beginning of heaven and earth.'

What then is the nature of a distinction? 'Distinction is perfect continence' (Spencer-Brown, 1972, p. 1) is the opening of Spencer-Brown's book. This sparse but very precise definition is characteristic of the flavour of the rest of the book. 'Continence' is derived from the Latin 'continere', meaning to contain, and the definition is saying that a distinction, i.e. the drawing of a boundary, perfectly separates that which is on one side from that which is on the other.

Out of the void we draw a distinction, and we can then separate that which is to be distinguished from everything else. Moreover, once it can be distinguished it can be indicated or identified. A distinction will only be drawn if there is some reason (intention or motive) for doing so, and there must therefore be some difference in value (to the person making the distinction) between the contents of the two states. We can give a name to the contents which then indicates the value. Saying or calling the name identifies the value, and implies the distinction. Thus the act of indication, at this almost pre-linguistic level, combines naming, acting and valuing all in one. It is reminiscent of Wittgenstein's (1958,

p. 3) proto-language where a builder shouts 'slab' to his labourer and the labourer brings one. The shout distinguishes the slab from other things, values it, and generates an action.

As a direct consequence of this definition, Spencer-Brown claims that two axioms can be stated – the law of calling and the law of crossing (see Figure 7.1).

Axiom 1 The law of calling: The value of a call made again is the value of the call.

Axiom 2 The law of crossing: The value of a crossing made again is not the value of the crossing.

For Spencer-Brown these capture the essence of a distinction – the difference between crossing and not crossing the boundary. The first axiom says that to draw a distinction and then to draw the *same* distinction again adds nothing new. To distinguish a circle and then distinguish a circle again leaves us with a circle. Thus, to re-call is to call. The second axiom involves us in crossing the boundary, and indicating the value by entering into the distinction. Now, if we draw another distinction, from within so to speak, we cross the boundary again and end up where we started with no indication. To distinguish a circle and then, from within, distinguish again must leave us with not-circle. Thus, to re-cross is to not cross.

To explain it a slightly different way. If we make a distinction we have a single circle distinguished against the blank background. If we then make a second distinction, relative to the first, then it can either go outside or inside the first distinction (assuming, since a distinction is defined to be 'perfect continence', that it cannot go across the boundary). If the mark goes in the same space as the first one – i.e. outside, then it is merely repeating the first one and so nothing is changed. If, however, it goes inside the first distinction then it is in a different space and, *from this space*, the second distinction can only take us back to the void.

One of the difficulties of grasping these concepts is that they are at a pre-linguistic level yet we inhabit language. As another example, consider a baby not yet able to talk. The baby cries (calls) for its mother's breast. This draws a distinction and values the contents. The distinction, the indication, and the value

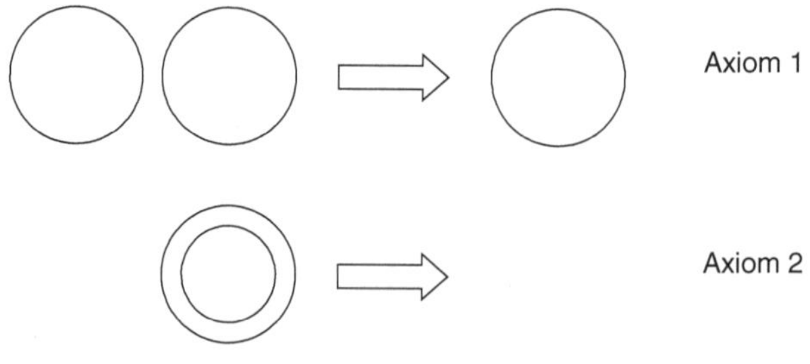

Figure 7.1 Spencer-Brown's axioms: the law of calling and the law of crossing.

are one in the cry. The baby cries again and again, each cry drawing the same distinction and re-calling the same call. This is Axiom 1. Then the mother arrives, the baby goes to the breast and the cries stop. The baby has entered into the distinction (mother's breast) and drawn another distinction (no longer need to cry for the breast) and peace returns. This is Axiom 2.

These two rather unintuitive axioms form the basis of the whole calculus of indications, as the laws of form are sometimes called. But what is the significance of this work?

First, all Boolean algebras rest on a group of assumptions, or a postulate set, which are stated without proof. All the theorems in Boolean algebra can then be rigorously deduced from this postulate set. There are numerous postulate sets for Boolean algebra (e.g. Sheffer 1913), all of which are essentially equivalent and none of which have previously been proved. Spencer-Brown takes Sheffer's set and shows that each postulate can be proved as a theorem in his algebra. Thus the whole of Boolean algebra (and its tremendously important applications in set theory and logic) can be shown to follow from the two axioms above and these, in turn, are a direct consequence of the primitive act of drawing a distinction.

Second, and potentially the most significant claim, is that the laws of form provide an important foundation for understanding human knowledge. The argument, with variations, runs that indication and distinction are essential elements in our perceptions and conceptions of the world and that the laws of form are therefore the laws of our description of the world and therefore our knowledge. Spencer-Brown thinks that mathematics is a very special subject in that mathematical forms represent and are derived from internal ways of thinking and that these have as much if not more validity than knowledge of external reality. Spencer-Brown sees the discipline of mathematics as a way 'of revealing our internal knowledge of the structure of the world' (Spencer-Brown 1972, p. xiii) and later suggests that we have a direct awareness of the mathematical form as an archetypal structure (Spencer-Brown 1972, p. xvi).

It is important that this is so because Spencer-Brown is attempting to explore this internal world. He believes that this is complementary to a study of the outer structure of reality, for what we approach – in either case, from one side or the other – is the common boundary between them, this boundary being the media through which we perceive the outside world. Thus the greatest significance of the laws of form lies in its explanation of this inner structure of knowledge in that it reveals the laws which must apply to our descriptions and understanding of the world, based, as they are, on distinction and indication. In this, Spencer-Brown is pursuing a similar task to the phenomenologists such as Husserl (1977, orig. 1931) and his ideas have been drawn on by Luhmann, who we will discuss shortly (section 7.4.3) (Luhmann 1996). Closer to real language, the laws of form can also represent propositional logic – Aristotelian syllogisms – and thereby underpin argument and deduction (Mingers 2013).

7.4.2 Concepts as difference and distinction

It is interesting to bring in here the other authors mentioned above.

Ferdinand de Saussure (1960, originally 1916) revolutionized the subject of linguistics and provided the foundations for structuralism as later developed by people such as Levi-Strauss (1963) and Chomsky (1957). Prior to Saussure, language was seen as primarily representational. That is, words were intrinsically associated with the objects they represented, be they actual objects or notional objects such as concepts and ideas. Moreover, there was a historical dimension in that if one traced back the roots of current languages and words one would expect to be able to see closer and closer links between the word and its object.

Against this, Saussure argued that in fact words bore very little relation to objects that they might stand for. Rather, a particular language is a complex system of differences and distinctions between terms – each language effectively cuts the world up in its own way. Particular terms or words then gain their meaning not directly from an object but only in terms of the system as a whole: only, in other words, through their similarities to and differences from other terms in the language. 'In language [*langue*] there are only differences. Even more important: a difference usually implies positive terms between which the difference is set up; but in language there are only differences without positive terms.' (Saussure 1960, orig. 1916, p. 120.) Saussure also brought in the important semiotic distinction between *signifier* and *signified*, already developed in a different way by Peirce (1931–1958). The word itself is a sign, a signifier, which goes with an associated idea, a signified, but which does not relate strongly to a real-world object. Moreover, in opposition to the historical approach to language, a signifier has no intrinsic relation to its signified – it is arbitrary in that it could be anything, although clearly once the association has been instantiated there is a de facto relationship. These ideas were taken further by Jakobson and Halle (1956) in terms of two dimensions of meaning – the *syntagmatic* (metonymy) and the *paradigmatic* (metaphor) (see Wilden 1977, Chapter 2). In the syntagmatic dimension, a term gains meaning through its combination with other terms in time or space. In the paradigmatic dimension, it gains meaning by its selection from other possibilities within a code or set of rules. For example, in the sentence 'fetch me the hammer', the word 'hammer' is combined with (syntagmatic) 'fetch me the ...' (it could have been 'where is the hammer?'), and selected from other possibilities (paradigmatic) such as 'screwdriver' or 'cup of tea'.

Gregory Bateson is not primarily a linguist but a cybernetician, although his work has been influential in many areas. In a seminal essay – 'Form, Substance and Difference' (Bateson 1973a) – he considers Korzybski's famous statement that 'the map is not the territory', in many ways the basic problem of epistemology.

> A map is not the territory it represents, but, if correct, it has a similar structure to the territory, which accounts for its usefulness.... If we reflect upon

> our languages, we find that at best they must be considered only as maps. A word is not the object it represents; and languages exhibit also this peculiar self-reflexiveness, that we can analyze languages by linguistic means.
>
> (Korzybski 1933, p. 58)

For Bateson, the question is what gets on to the map? And the answer he gives is *differences*. The world, the territory, which of course we can never *directly* access, cannot be simply uniform, for then nothing would get on the map. It must, at some level, consist of differences in temperature, pressure, height, vegetation etc., and it is some selection of these that get transformed to become the map. The same is true in the process of perception and cognition. Differences in the environment are 'picked up' subconsciously by the nervous system and transformed into that which we experience. The interesting questions are: what gets picked up out of the infinite differences that exist? And what do they become transformed into? Bateson called this the 'difference that makes a difference' and identified it with the basic unit of information.

Neurophysiologically, we can see that the nervous system is organizationally closed and structurally determined (Maturana 1980a). This means that the differences that get 'noticed', that 'make a difference', are always determined by the structure and state of the nervous system itself from instant to instant, and that the effect that is generated will also be structurally determined. We must not forget, of course, that in evolutionary terms the structure of the nervous system itself develops in such a way as to enable the continuation of autopoiesis through its structural coupling with its environment.

7.4.3 The boundaries of language

In the previous section we considered the nature of individual concepts and the extent to which they were distinct and therefore bounded. But we can also look at the level of languages or forms of communication as a whole.

Consider first Luhmann, who we discussed in Chapter 6, for whom the basic social act is in fact *communication* (Luhmann 1986): that is, an utterance or speech act which carries information and generates understanding (and thereby further action) in another person. Society is thus a network of communications continually generating further communications. Modern societies have become structurally complex, embodying a range of functionally different subsystems such as the economy, law, science, politics or the mass media. Luhmann suggests that this demarcation occurs precisely through a differentiation of types of communications. The general, everyday communications of the lifeworld differentiate themselves into specific subsets which have implicit closure, i.e. boundaries. Different subsystems, whether they are societal ones such as those mentioned above, or other social systems such as organizations, generate closure at two levels.

First, operations always generate other operations of the same type – thus communications generate communications (social systems), decisions generate decisions (organizations), and thoughts generate thoughts (psychical systems). There are connections between the different domains – what Luhmann calls interpenetration – where a particular element, e.g. a meaning, is part of both systems but it is not passed from one system to the other. The second form of closure is the type of communication. Thus a particular subsystem, say law, will only recognize or accept communications that are appropriate for law and those which are accepted will then generate further 'legal' communications. The distinction (see the discussion of the laws of form above) as to what does or does not belong to the legal system is made by the system itself – it draws its own boundaries and is thus self-referential. These subsystems do interact with other subsystems but only in the autopoietic sense of triggering responses. Thus a legal communication about a fine may trigger an economic communication for payment.

We can also consider the boundedness of spoken languages such as English. A competent speaker of a language automatically generates sentences that are in that language. The sentence could be grammatically incorrect, but it would still clearly be English rather than French. This is because there must be some underlying generative mechanism (Chomsky 1957) which automatically produces sentences in a language, and at any particular time a person is actually speaking only one language. Bilinguals can of course switch from one language to another, but they are always either within one or the other. Indeed, as will be discussed in Section 8.5.3, speaking a different language brings with it much more than just the language – it also involves gestures, emotionality and a whole way of being (Wittgenstein 1958).

Even here, however, the situation is not always that clear. For instance, if you listen to a group of people from different countries talking, they may use one language, say English, as a lingua franca but then bring in parts of other languages as they go, producing a spontaneous and ill-defined polyglot.[7]

7.4.4 Concepts, language and boundaries

What do these theories of difference, distinction and language tell us about the question of boundaries?

Taking first the laws of form: unsurprisingly perhaps, given its relationship to mathematics, it makes the same fundamental assumption that distinctions are 'perfect', i.e. that the boundary created by a distinction is unambiguous and impermeable. But, as discussed in Section 7.3, while this may work in the conceptual world it does not when we try to apply it to the real world of objects, or indeed social categories. If we take the physical world it seems that there is a valid distinction between 'table' and 'not table', although even here the distinction may not be so clear at the atomic or subatomic level. When we come to the social world and such concepts as good/bad, dominant/submissive, competitive/cooperative, can we really say that each of these distinctions is crystal clear and that people can be assigned to the different categories without problem?

We can consider this in terms of Frege's distinction between sense and reference (Frege 1952). Words, or indeed concepts, have a sense or meaning, and they also have a reference – i.e. that to which they can apply or refer. So the 'morning star' and the 'evening star' have different senses or meanings even though they both refer to the same object – Venus. A concept such as 'terrorist' is, in Spencer-Brown's sense, a distinction, which is the equivalent of its meaning or sense.[8] However, it is clearly a rather fuzzy term and it would be difficult to draw absolutely clear boundaries of what activities would constitute terrorism. And from Sassure's perspective any such meaning would come about not as a positive or even ostensive definition but through its differences to other, similar, terms such as guerilla, mercenary, soldier or freedom fighter.

In terms of its reference, i.e. whether or not it can be applied to a particular person, there is even more of a problem as Checkland (1983) points out: one person's terrorist is another person's freedom fighter. So, regardless of how precise or imprecise the term (i.e. the boundary) is, the application of the term always assumes a particular perspective or worldview.

Framing the discussion in terms of differences, rather than clear distinctions, does not seem to me to help for there is a mutually recursive relationship between the two. To speak of a difference implies that there must be a distinction such that there are two distinguished states between which there is a difference. So, the distinction comes before the difference. On the other hand, how do distinctions come to be made without there being differences? – as Bateson says, differences that make a difference, i.e. create a distinction.

Where does this leave us? I think we have to accept both that concepts, and language generally, are intrinsically fuzzy and that the application of concepts to referents, especially in the social world, will always be based on perspective and values. However, in a way this is no different from what we have discovered with physical boundaries. They too are not perfect and have degrees of fuzziness and impermeability about them, but this does not stop them acting as boundaries so long as they have the requisite degree relative to the role they are playing. The same is true with language – concepts need only be clear and applicable enough for language to 'work', for us to be able to achieve mutual understanding and successful co-ordinations of action. That we do do this is manifestly the case. When language breaks down, or when there is disagreement about meanings, then we should be able to follow Habermas's theory of discourse and initiate questioning and debate about implicit validity claims.

The fact that concepts are intrinsically relative and debatable does not make the whole matter purely arbitrary. In the same way that critical realism accepts the epistemic relativity of knowledge (Section 2.3.2) without thereby saying that all theories are equally valid, so we can make judgements about the more or less appropriate use of concepts. For example, Bhaskar (1994, p. 110) points out that genocide such as that carried out by the Nazis could be described as 'depopulation', 'people dying', 'people being killed', or 'people being murdered'. All of these are true, but the last one is the most precise and accurate – none of the others carry the full perlocutionary force and they are therefore less correct.

Moreover, this is not a matter of personal perspective or value but of descriptive and explanatory adequacy. We should not confuse the map with the territory, but nevertheless the map is a guide to the territory and may be a better or worse one.

7.5 The boundaries of social systems

The question of the boundaries of social systems, or even the existence of social systems, is extremely contentious because of major debates about the nature of the social world. The whole issue of whether society can be seen as a social system was the subject of Chapter 6, and the specific problems of drawing the boundaries of social systems was covered in Section 6.4.3, so we will not repeat that material here. Suffice to say that Tilly's (2004) work on social boundary mechanisms resonates will with the mechanistic approach of this book.

7.6 Self-bounding through organization closure

In this section I wish to pick up the discussion in Section 7.2.3 about natural wholes. In that discussion we were primarily concerned with physical systems, particularly those occupying a particular area of space-time, and the extent to which they form boundaries independent of an observer. I want to extend that discussion to consider more complex systems which include non-physical components for which boundary identification is much more problematic. One approach is to consider the extent to which systems form their own bounds, or at least demarcate themselves from their environment, through some kind of organizational closure (in Maturana and Varela's terminology) or self-reference.

There is a more general notion propounded by Maturana and Varela, that of organizational closure, of which autopoiesis is a special case (Maturana and Varela 1980; Varela 1977, 1979). Mingers (1997b) argued that organizational closure simply requires some form of self-reference, whether material, linguistic, logical, or social, rather than the more specific process of self-production. However, as Teubner (1993, p. 16) has noted, there is considerable conceptual confusion around such terms in the general systems literature as well as within philosophy (Bartlett and Suber 1987) and even, on occasions, within Maturana and Varela's work. What is the difference between, for example, 'self-reference', 'self-production', 'self-organization', 'self-regulation', 'self-observation', 'circular causality', 'tautologies and contradictions', or 'auto-catalysis'? While not claiming a complete analysis of such systems, I offer the categorization in Table 7.1 as a start.

In this typology, the numerical 'Boulding level' refers to Boulding's (1956) Hierarchy of Systems as modified by Mingers (1997b). The primary feature that distinguishes the different levels is the type of relation that emerges at each level; systems at a particular level involve the relations of that level plus those from preceding levels.

- Level 1 systems are characterized by static relations of topology;

Table 7.1 Typology of organizationally closed (self-referential) systems

Boulding level	*Type*	*Characteristic*	*Example*
1, 2	*Self-influencing* systems	Circular causality, causal loops	Size and birth rate of population
3	*Self-regulating* systems	Maintenance of a particular state variable	Thermostat, body temperature regulators
3	*Self-sustaining* systems	Organizational closure but not self-production	Gas pilot light, auto-catalysis
4, 5	*Self-producing* systems	Autopoiesis	Biological cell, the game of Nomic (?)
6	*Self-referential* or *self-referring* systems	Symbolic reference to self (e.g., linguistic or pictorial)	'This is a sentence', Escher's *Drawing Hands*, Magritte's *The Treachery of Images*
7	*Self-conscious* systems	Able to interact with descriptions of self	A person saying: 'I acted selfishly today'

- level 2 by dynamic relations of order;
- level 3 by feedback relations;
- level 4 by relations of self-production;
- level 5 by relations of structural coupling between components;
- level 6 by symbolic relations;
- level 7 by relations of self-awareness.

The first type I identify as *self-influencing* systems. These are examples of what are often called causal loops or circular causality, that is, patterns of causation or influence which become circular: for instance, the larger a population, the greater the number of births, and thus the larger the population. This is a positive loop leading to exponential increase or decrease; more commonly there are negative loops which lead, at least temporarily, to stability, for example the price/demand relations for a normal good. The economic and ecological domains are full of complex patterns of just such mutual causality. The second type of system that I would distinguish are *self-regulating* systems that are organized so as to keep some essential variable(s) within particular limits. They rely on negative feedback and specified limits. Obvious examples are a thermostat and the blood temperature control system. These are distinct from the self-influencing systems in that they maintain a pre-specified level determined by the wider system of which they are a part. Self-influencing systems may stabilize through negative feedback but do so at essentially arbitrary levels.

The next type I call *self-sustaining* systems. In Maturana and Varela's terms these systems are *organizationally closed* but not *self-producing*. Their components and processes close in upon themselves so that they themselves are both necessary and sufficient for their own continuance. Such systems do not however produce their own components. A good example is the gas heater pilot light found on many central heating systems. In such a system, the gas pilot light heats a thermocouple which controls the flow of gas to the pilot light which allows the pilot light to function in the first place. Once it is in operation, it sustains itself. However, once it stops it cannot restart itself – it needs some form of outside intervention to begin the cycle again. Other examples are systems of auto-catalysis where a chemical reaction produces at least some of the chemicals that are necessary for the reaction to occur; and the nervous system where every state of nervous activity is triggered by a previous state and triggers, in turn, the next state.

At the next level we have autopoietic systems which are *self-producing* of both their components and their own boundary. They are more than self-sustaining in that they actually produce the components necessary for their own continuation. Such systems have properties such as autonomy, since they depend mainly on their own self-production, and identity, since they maintain their own individual autopoietic organization despite changes in their structure. Next come *self-referential* or *self-referring* systems. To avoid confusion, I restrict these specifically to *symbolic* systems which allow self-reference, and particular instances of this, rather than individual organisms referring to themselves, which come in the next category. Such self-referential examples will generally be linguistic or pictorial and include all the many paradoxes and tautologies (Hughes and Brecht 1978): for example, 'This sentence is in English', or 'This assertion is not true'. Pictorial examples are Escher's drawing (*Drawing Hands*) of two hands drawing each other, and Magritte's pictures, one (*The Treachery of Images*) of a pipe with the words (in French) 'this is not a pipe', and another (*The Human Condition*) in which the picture contains a picture of part of the scene.

Finally in this typology, there are *self-conscious* systems that can, through language, create descriptions of themselves and then interact with these descriptions, thus recursively generating their conscious selves. Human beings are the main example at this point in time. This occurs mainly through linguistic utterances, either latent or manifest, such as, 'I am lying'; or 'I hereby promise to …', but can also be embodied in gestures of facial expressions. Note that I have not included Boulding's level 8, social systems, because of the extensive debate about social systems covered in Chapter 6.

The general point of this section is that all these types of systems generate, through their own operations, a form of closure which is in essence a type of boundary. This is not necessarily a physical boundary, or one marked by particular boundary components, but it is nevertheless a self-generated, and therefore observer-independent, demarcation of the system from its environment.

7.7 The problem of the observer

So far we have largely ignored the problem of the observer, simply discussing the characteristics of different types of boundaries as if they were observer-independent. However, we must now debate this issue.

In the early days of systems science, still under the thrall of empiricism, systems and their boundaries were assumed unproblematically to exist independently of the observer. Indeed this was the hallmark of positivist science – the elimination of any degree of subjectivity (Mingers 2000a). It is still the case today in some areas of general systems theory and hard systems. However, from a relatively early time systems thinking did begin to recognize the involvement of the observer in determining systems and their boundaries. Beer, for example, states this very clearly in one of his early works:

> [T]he situation becomes still more confused when one stops thinking about physical objects and considers systems instead. The mouse and the plant may be physical objects but they can be understood only as systems. Moreover, the boundaries of these systems are not the same as the boundaries of the physical objects themselves.… At this point the scientist joins the philosopher in his assault on the classical notion of science: to measure aspects of a system – to observe it even – is to alter the system so measured and observed.
>
> (Beer 1966, p. 96)

> A system is not something given in nature, but something defined by intelligence.... We select, from an infinite number of relations between things, a set which, because of coherence and pattern and purpose, permits an interpretation of what otherwise might be a meaningless cavalcade of arbitrary events. It follows that the detection of system in the world outside ourselves is a subjective matter. Two people will not necessarily agree on the existence, or nature, or boundaries of any systems so detected.
>
> (Beer 1966, pp. 242–243)

It must be said that having made this strong assertion, his book reads for the most part as though systems *do* exist unambiguously in the world; there is little discussion of the difficulties that this stance engenders. Organizations, for example, are taken to *be* complex systems even though there could be much debate about their boundaries.

7.7.1 System boundaries as observer constructs

Two other systems thinkers took this stance more thoroughly and between them completely re-oriented the disciplines of systems engineering and operational research (Peter Checkland) and cybernetics (Humberto Maturana) to an interpretive or constructivist perspective. The move from first-order cybernetics (of the observed) to second-order cybernetics (of the observer) has been well documented by Hayles (1999). We have already described this movement in Sections 1.2.1 and 3.2 and so we will not repeat this other than to note that, for Checkland, boundaries are a distinction made by an observer marking something *taken to be* a system. We cannot assume that such systems and boundaries actually exist in the external world. For an extended explanation and critique, see Mingers (1984). And for Maturana we cannot assume that our perceptions correctly represent some independent environment. Nor, in neurophysiological terms, can we distinguish between reality and illusion. As we cannot access an independent reality, we should suspend our naive belief in it. Instead, we should acknowledge that existence for us is constituted by us through our linguistic distinctions. This is what is meant by 'constitutive ontology' – we can only interact through linguistic entities and they come into being as they are constituted by us.

Thus, although coming from different directions, sociology for Checkland, biology for Maturana, the two perspectives converge to a similar point at which the external world has become completely inaccessible and we are confined within the domain of the constructions of the observer. Not just boundaries, but systems and perhaps even objects are denied ontological existence.

I would argue that in many ways this undermines the force of systems thinking. Systems thinking began (in modern times) with the cyberneticians of the 1930s who found the concepts *necessary* to explain puzzling features of the world. The way in which organisms could display complex and apparently purposeful behaviour with no central control led to the concepts of negative feedback and information; the cyclical patterns of equilibriating and disequilibriating behaviour that occurred in so many different domains led to the notions of interacting positive and negative feedback loops; and the failure of reductionist thinking to explain the diversity and persistence of the biological world led to ideas of holism and emergence. These were more than mere epistemological devices to organize our thinking, they were genuine explanatory concepts in that the existence of such systemic processes in the world was *necessary* to explain the phenomena that were observed. To deny reality to systems concepts is to reduce them to an essentially arbitrary language game.[9]

There is not space here to make these arguments fully (see Mingers 2000a) but I will summarize them briefly. Checkland is right to recognize that we do not have access to the world in a pure, unmediated way. Clearly, as human beings we can only ever experience anything through our perceptual and linguistic apparatus. It does not follow from that, however, either that our descriptions are unrelated to the world or that we should deny existence to anything simply because our knowledge or perception are limited. This is to commit the *epistemic fallacy* (Bhaskar 1978): that is, believing that statements about *being* can be analysed or limited by statements about our *knowledge*. Checkland is also right that we can never know definitively or prove conclusively the existence of systems. Again, however, this does not prove the converse, that they do *not* exist. We can move beyond the crude empiricist ontological criterion that *to be is to be perceived*, and instead adopt the critical realist view that causal efficacy is the proper criterion for existence. In other words, if some structure or system can be shown to have causal effects on the world then, whether we can perceive it or not, it can be said, putatively, to exist. Given this criterion, we can take particular phenomena that we wish to explain, hypothesize possible generative mechanisms *which, if they existed, would generate the experienced phenomena*, and then attempt to confirm or refute them. This philosophical stance grants possible reality to both physical and conceptual systems while recognizing the inevitable observer-dependence of our descriptions, and allowing that the social world is inherently different from the natural world.

7.7.2 System boundaries as process

I want now to consider a more recent contribution by Midgley (2000). In general, I am very much in agreement with much of the thrust of Midgley's work, not least the fact that he gives the idea of boundaries a central position. Midgley's primary interest is in methodology for systemic intervention, but he sees that any intervention will require definitions or judgements to be made about the boundaries of the system to be considered.

He identifies the primary problem with boundaries as the implied separation between observer and observed and in particular the traditional empiricist view that there could be objective, observer-independent observation. Midgley reviews four systems approaches, each of which tries to resolve this problem: von Bertalanffy (1971), Bateson (1973b), Maturana (1980), and Fuenmayor (1991). They are all found wanting: 'while each of these authors has managed to find an alternative to the radical separation of observer and observed, a less naïve subject/object dualism is nevertheless still evident in their work' (Midgley 2000, p. 67).

To resolve these problems Midgley moves to what he calls (after Whitehead) a process philosophy. By this he means one that moves away from specifying particular content or boundaries to one that places primacy on the *process* of generating boundaries. I shall trace his arguments in some detail as I think it is an original, if ultimately flawed, approach.

Von Bertalanffy, Maturana and Bateson, in different ways, all specify particular types of systems as drawing boundaries (or more generally, for Midgley, creating knowledge) – open systems, autopoietic systems, systems of transmitted differences. Midgley wants to move away from privileging any one type of content to the *process* of specifying that content.

> [P]rocess philosophy involves identifying a process that is not dependent on further identification of a single type of system giving rise to that process.... If we regard the process of making boundary judgements as analytically prime, rather than a particular kind of knowledge generating system, then subjects come to be defined in exactly the same way as objects – by a boundary judgement.
>
> (Midgley 2000, p. 79)

Midgley suggests that we need to consider both outward-looking, first-order boundary judgements about the world and backward-looking, second-order judgements *about the system making the first-order distinctions*.[10] When we make this second step we can pick out many different possible observing systems and have many different theories about how these systems generate knowledge. All of these systems must, however, involve what Midgley calls 'sentient beings' (human or animal). This in fact means that there must be two second-order judgements – one concerning the boundaries, i.e. the extent of the knowledge-generating system, and one concerning its nature. These judgements are not interest-free but depend on the purposes, values, and knowledge available at the time. As these change, so the second-order boundary judgements can change. Moreover, we can (and perhaps should?) examine the system *that is making the second-order judgements* and perform second-order boundary judgements on that, and so on ad infinitum. In other words, we quickly end up with a plethora of potential systems: (i) the first-order system boundaries that are distinguished 'in the world'; (ii) the different second-order system boundaries that may be constructing those first-order systems; (iii) the different 'third-order'[11] systems that are distinguishing the second-order ones.

Midgley does not develop this in more detail on the grounds that to do so would be moving into specifying content. Thus he does not define what he means by 'sentient beings'; nor does he elaborate on how boundaries are in general specified, although he does discuss certain particular instances, for example multiple stakeholders within organizations.

> Content philosophy presents a theory of exactly what counts as a knowledge generating system, while process philosophy allows for a variety of possible knowledge generating systems (with the proviso there are sentient beings identified as part of them). Also, content philosophy is mono-theoretical ... while process philosophy allows for theoretical pluralism.... The reader may be left wondering why I have only talked in broad terms about the process of making boundary judgements.... The answer is that, as soon as we move away from discussing boundaries in general to a generative mechanism, we have moved away from process to content!
>
> (Midgley 2000, pp. 88–89)

Finally, and crucially, Midgley elucidates what his process view means we can, or cannot, say about 'reality'. He discusses this in relation to three major philosophical paradigms – realism, idealism and social constructivism.[12] His approach is to show that each paradigm makes particular, and different, judgements (often implicit) about the kinds of boundaries that can be drawn and the proper kinds of knowledge-generating systems. He then goes on to argue that *each* of these paradigms is *compatible* with a process view in the sense that it can be explained in terms of first- and second-order boundary judgements. He does not see process philosophy as falsifying any of these paradigms, or as subsuming them; all it does is to allow us to avoid 'slipping into a dogmatic insistence (sometimes found within these paradigms) that there is only one correct boundary to work with' (Midgley 2000, p. 98).

In evaluating Midgley's contribution I observe several problems.

- He is concerned to avoid the *ontological* dualism between subject and object and claims to replace it with an *analytical* dualism between first- and second-order judgements. He also claims to steadfastly avoid making assertions of content. I do not accept either of these claims. We have, in fact, just the traditional distinction between observed and observing systems (Von Foerster 1984), first- and second-order observers (Luhmann 1987a) or systems as epistemology not ontology (Checkland 1989) cast in a different form. With all these earlier attempts to refrain from ontological commitment and place priority on one side of the distinction only (the observer's), there are always ontological assumptions smuggled in. It is just impossible to theorize about anything without there being a 'thing' to theorize about.

  Midgley's approach is different in not privileging one side of the distinction, but it still cannot avoid ontological commitments. If boundaries are being drawn by some system, first or second order, then clearly a system *exists* (or must be assumed to exist) that is drawing the boundaries; the boundaries so drawn, in that they can have causal effect, must also have existence. Equally, Midgley's scheme only works if we assume the existence of a 'real world' on which first-order boundaries can be drawn, and some other, unnamed, system that is capable of drawing second-order distinctions.

Moreover, Midgley clearly does specify content when he says that second-order systems must include 'sentient beings' – being evasive about what might count as such is merely avoiding the issue, even more so when he says second-order judgments must include determining the nature as well as the extent of such beings.

- Midgley's critique of existing philosophies is based on their exclusivity and commitment to a single theoretical paradigm. In the case of realism this is certainly not true of *critical* realism. As we have seen, CR is highly pluralistic both in its ontological acceptance of many different kinds of entities – physical, social, conceptual, or moral – and in its epistemological acceptance of a necessary range of different research methods from quantitative to qualitative. CR also takes as fundamental that there is an intimate connection between fact and value – something that Midgley espouses but presumes that realist philosophies do not.
- The whole 'process' seems quite unhelpful in practice. If carried out properly it would lead immediately to a massive proliferation of systems and systems boundaries through an infinite regress of different orders of boundary judgements. Although Midgley does not actually propose such a comprehensive review, nor does he deny it. In the later practical chapters he is quite selective in doing this in particular contexts, but little general guidance is offered (for this would presumably be content) as to how better or worse boundaries might be drawn (or are they all equally valid?) in any particular situation, or how the whole process could be effectively grounded in order to prevent infinite regress.
- In overall terms Midgley seems to have gone to great lengths to avoid making commitments and has thereby thrown out the baby with the bathwater by remaining silent on the most fundamental issue – the relationship between the real world and the boundaries we draw about it. Midgley would no doubt see this as an unwelcome return to subject/object dualism, but I make no apologies for that. The question that is never really answered is this: are all boundaries *simply* constructs of the observer (the second-order knowledge-generating system), i.e. independent of any world external to the observer?

  If the answer is 'yes', then we remain in the subjectivist world of Checkland and Maturana. Surely this cannot fit in with Midgley's whole enterprise, which aims at *intervention* into the world. Given this agenda then surely he must accept that there is such a world; that the boundaries we draw have effects on that world; and that we can therefore judge boundaries as being better or worse from some point of view.

  If the answer is 'no', then there must be some relationship or interaction between the boundaries that are drawn and the world that they relate to. The implication is then that we must have a view about the nature of that relationship and particularly how boundaries drawn by observers relate to boundaries that may exist in the world.

7.7.3 Summary

We have explored in this section a significant challenge to the ontological reality of systems boundaries. Both Checkland and Maturana have arrived at similar points, albeit from different directions. For them, systems thinking is a question of epistemology rather than ontology; systems boundaries are distinctions brought into being by an observer and cannot be related directly to some external reality. Indeed, for Maturana such distinctions are themselves constitutive of reality. Midgley argues a slightly different point: that we must focus on the processes by which boundaries are constructed rather than the nature of the boundaries themselves. Ulrich (Ulrich 1994, 2000) is also someone who has placed boundaries, or at least boundary critique, at the heart of his work. He fits in broadly with the writers in this section, as he sees boundaries very much as the result of decisions or judgements made by systems inquirers or designers.

I have argued, against both these views, that system boundaries can and sometimes do exist in the world independent of human observers. The basic argument is the critical realist one that if we can show that boundaries are causally active then they must exist. Put another way, if we find it necessary to postulate the existence of a boundary in order to explain a particular phenomenon then we are entitled to accept its existence. This does not only hold for physical boundaries but potentially also for conceptual, psychological, or social ones. This argument does not mean that it is easy to identify actual boundaries in practice, or that we can ever be *certain* about them. In our work as systems theorists or practitioners we always remain in the transitive domain. Our systems models themselves are just that, humanly-constructed models, maps not to be confused with the territory. But as models, they can be said to be models *of something* intransitive, something different from themselves. And we can then discuss and investigate their relationship to that something.

7.8 Boundary setting in practical interventions

So far we have been primarily concerned with recognizing or determining the boundaries of systems that already exist in some way. But much of *applied* systems thinking is concerned with undertaking projects or interventions in the real world in order to bring about change that is in some sense desirable for particular actors. In these types of situation we are often in the role of setting rather than observing boundaries, or perhaps being the subject of boundaries that have been set by others. In a general sense we can classify these boundaries as relating to:

- Boundaries that delimit the scope of the problem. They could be literally spatial, or at least geographical, as in which regions or areas are to be included. They define what problems or aspects of the problem situation are to be included (in 'The Strategic Choice Approach', 2001, Friend discusses this under uncertainty about related problems). They also determine what range of stakeholders are to be included either as clients, problems, owners or perhaps victims, and which are to be excluded (Midgley 2000 refers to these as the sacred and profane). The scope is strongly related to levels within an organization – the greater the scope of the problem the more, and higher, levels that are likely to be involved (Beer's 1985 Viable System Model requires the specification of a 'system in focus' and consideration of the levels directly above and below).

- Boundaries that delimit the time horizon to be considered. The shorter the time period, the more things that will be fixed or unchangeable and therefore have to be taken as given. Generally, therefore, the shorter the time scale the more narrowly will be drawn the scoping boundaries. Note that terms such as 'short term' or 'long term' are not absolute but relative to the nature of the situation. In the aircraft industry, five years is short term; in a bakery, a day is.
- Boundaries that place limits on what may be changed or what must be taken for granted. Any organizational setting will be sedimented with cultural norms, ways of working, and unquestioned assumptions about the nature of the situation (Mingers 2000c refers to questioning rhetoric, tradition, objectivity, and authority; and Ulrich's 2000 critical systems heuristics can be used to challenge expert judgements). These will often never be made explicit and yet they will powerfully shape the extent of potential changes. Indeed, one could see the process of bringing about change as essentially one of overcoming or transgressing boundaries (Mingers 1997c).

7.9 Conclusions

This has been a long and complex chapter which has tried to reconsider the whole question of the nature of boundaries within systems thinking – a long-neglected question. I am not sure that we have 'solved' the problem of boundaries but I hope we have addressed the important issues.

We began by considering, from an almost phenomenological view, what features of boundaries came to light when we looked at relatively simple physical systems. The main points I would wish to bring forward are:

- Boundaries do exist independently of an observer, although it is always an observer who chooses to observe them. Boundaries are of different types: edges and surfaces, enclosures and membranes, and demarcations; and they have particular effects – separation, containment, and constraint.
- Boundaries are never 'perfect' but always to some degree fuzzy and permeable. This depends on the space-time frame with which we observe them. This does not mean, however, that they do not function as boundaries – they only need to be 'good enough' relative to the role that they play.
- In picking out or 'presencing' particular boundaries relative to some purpose, observers do not simply *perceive* systems but *conceive* them. That is, they are selective in the boundaries they draw dependent on that which

they wish to understand or explain. Moreover, there may be multiple possible boundaries around a set of components, and a component may be a member of multiple systems.

From physical boundaries we moved to consider mathematical and conceptual boundaries:

- Geometry provides a formalism for considering boundaries, particularly spatial ones, but its assumptions are too idealized to apply directly to the real world. Set theory introduces the idea that elements can be separated without an actual boundary at all by a membership criterion. Moreover, operators can lead to entry or exit, i.e. input or output; a crossing or not crossing of the notional boundary. As in geometry, set theory assumes a perfectly distinguished boundary.
- Concepts and language are similar to mathematics in the sense that underlying them is the idea of a perfect distinction, i.e. a completely clear demarcation of a concept from any others. Again, in practice this does not hold, with distinctions and differences being both imprecise and inevitably judgemental in their application. But, as with physical boundaries, distinctions do not have to be perfect to work and to allow us to communicate and interact.

After briefly considering social systems we moved to discuss a central cleavage within systems thinking – whether boundaries could be said to exist at all, ontologically, or whether they were in fact always simply constructs of the observer. Without replaying the debate, my conclusions basically followed critical realism. We have to first distinguish between the transitive and intransitive domains. In the transitive domain we are always dealing with our own human constructions and models and we, as humans, can never escape this domain. However, we can use the criterion of causal efficacy to argue that we can take boundaries to exist in the intransitive domain if we find it necessary to postulate them in order to explain our experiences. This is a powerful argument for the independent existence of both systems and their boundaries.

Notes

1 Very precise definitions of such things can be found in geometry and topology – see section 7.3.
2 Euclidian space is an n-dimensional space of real numbers in which distances between points are measured using Pythagoras's Theorom.
3 Note that a sphere has two sides (inside and outside) but no edges. You cannot move from outside to inside without going through.
4 There are many websites about laws of form – see www.lawsofform.org/.
5 The laws of form have close connections to C.S. Peirce's entitative graphs – see Roberts (1973).
6 Briefly, the distinction between these is that an arithmetic uses constants whose values are known (e.g. 2, 5, etc.) whereas an algebra is concerned only with those properties of an arithmetic which hold irrespective of particular values, e.g. $a^2-b^2=(a-b)(a+b)$.

7 The term 'lingua franca' is actually the name of just such a polyglot combination: a mixture of Italian with Provençal, French, Spanish, Arabic, Greek, and Turkish, formerly spoken on the eastern Mediterranean coast. Another example is 'franglais' – French with added English.

8 An intensive definition in terms of set theory.

9 In response, Checkland has said (in a private communication):

> in my experience it is not a case of Hard ST or Soft ST as you imply but Softest/Hardest with Hard being the occasional special case of Soft. Usually I find myself working with various models with different W's; but occasionally it is fruitful and not harmful to choose to see a particular bit of the world as 'a system' and use HST. Operating with SST subsumes HST with the latter being a *conscious* choice.

This does not seem to me to address the main argument as it implies choice rather than necessity.

10 Midgley acknowledges that this distinction is similar to that of first- and second-order cybernetics. His approach differs in not privileging either the first- or second-order view, but seeing both as an equal part of one process.

11 My suggestion.

12 There are many ways of classifying different paradigms and I do have some problems with Midgley's, which I will discuss below.

8 Knowledge and forms of truth

8.1 Introduction

It can be argued that the standoff between positivism and interpretivism has been ameliorated (Mingers 2004b) in favour of some form of pluralism, either one which simply accepts the validity of different paradigms, e.g. Robey (1996) or Jackson (2000), or one which seeks actively to combine research approaches, e.g. Goles (Goles and Hirschheim 2000), Tashakkori (Tashakkori and Teddlie 1998), or Mingers (2001a). However, where does this leave the question of knowledge? Are there different forms of knowledge depending on the paradigm in use? And how does this relate to truth, which is supposedly an essential characteristic of knowledge as opposed to mere belief?

This chapter will address these issues from the perspective of critical realism, which, as we have seen, accepts aspects of both positivism and interpretivism but maintains a strongly realist and critical core. We will contextualize the argument by considering a recent debate within the management literature between Meckler and Baillie (2003a; 2003b) and Gioia (and others) (Gioia 2003; Lounsbury 2003; Ryan 2005), and particularly a response by Hunt (2005). Meckler and Baillie proposed a 'middle way' between strong positivism and strong constructionism based, in part, on a form of correspondence theory and Searle's (1996) set of distinctions between epistemic and ontological objectivity and subjectivity. Gioia strongly attacked this as simply a way of assimilating interpretivism to positivism. Hunt essentially backed Meckler and Baillie by arguing the case for scientific realism (as distinct from critical realism mentioned above) and a particular conception of truth intimately related to trust.

The argument in this chapter is in broad agreement with Hunt except that I would wish to maintain a more polyvalent view of truth, and indeed knowledge. To put it simply, I accept the ontological claims of realism for the existence of a subject-independent and causally efficacious world. But, I argue, within this world there are substantively different kinds of things that can be the objects of knowledge to which we have different forms of epistemological access. Thus there is neither one kind of knowledge nor one kind of truth.

The chapter begins with a brief review of the Meckler and Baillie debate and particularly Hunt's response. It then moves to review conceptions of knowledge within knowledge management and points out their limitations, not least the lack of connection to truth. The next section describes various theories of truth, focusing on ideas from critical realism and the recent work of critical theorist Jürgen Habermas (2003b). Following this, I show that there are many ways in which we validly talk of 'knowing' something and I identify some dimensions which underlie all of them. One of these dimensions is that of truth or, more generally, warrantability or justification. This is essential in distinguishing knowledge from mere belief or assertion. I then bring these threads together to present a typology of four distinctively different forms of knowledge, each with different possibilities of truth or warrantability. Finally I consider the implications of this theory for both knowledge management and management knowledge.

8.2 The realism debate

Meckler and Baillie's (2003a) project was to develop a position in between constructionism (or postmodernism as they sometimes, perhaps incorrectly, called it) and positivism. They wanted to maintain the notion of truth as a broad correspondence between statement or beliefs about the world and the way the world 'actually is' without accepting some direct, positivistic relation between sentences and facts. In this, they are consonant with the critical realist view. Their main aim, and most of their paper, was actually directed against what they took to be the constructionist view (Astley 1985) that there could be no objective truth since the product of social research was only ever further concepts and theories, never statements that could be made true (or false) by some external reference. Truth was always socially constructed. Meckler and Baillie did accept that the world of social facts and events was humanly constructed but drew on Searle's (Searle 1996) work to argue that the social world was ontologically subjective but still potentially epistemically objective.

Gioia (2003), in a vituperative response, sees it all as a plot to assimilate constructionism back into the positivist fold. The central disagreement can be described quite simply. For Meckler and Baillie there is a reality of facts and events (accepting the differences between the physical and social), and there is a humanly-constructed world of beliefs and theories which may be true or false in relation to that reality. Gioia accepts that things do exist and events do occur, there are indeed 'facts', but holds that these in themselves are uninteresting or perhaps irrelevant until they become part of the social world through interpretation, discussion and debate. For Gioia it is this world of belief and interpretation that is real, not the facts in themselves. 'The actions, events, observations, and so forth might be common, but those are mere data from which the interpreted world is assembled.... The world we deal with is the interpreted world, not the world rendered in objective facts.' (Gioia 2003, p. 287.) Reality, or at least *social* reality, is only that which results from peoples' interpretations, and therefore truth must be bound to that reality as well. It is either the relative truth of different interpretations and valuations, or, on occasion, 'an intersubjectively agreed truth arrived at by negotiated consensus' (Gioia 2003, p. 288). I will discuss this later in terms of Habermas's consensus theory of truth.

Hunt (2005) is generally sympathetic to Meckler and Baillie's argument but feels that their Tarskian theory of truth is impoverished, and so he presents a more sophisticated version based on scientific realism (SR). For Hunt, SR is based on four principles – (i) classical realism, which maintains that the world exists independently of its being perceived or represented; (ii) fallibilistic realism, which accepts that knowledge can *never* be known with certainty – it will always be fallible; (iii) critical realism,[1] which maintains that because of ii) we must always critically evaluate our theories; and (iv) inductive realism, which holds that the long-run success of a theory does provide evidence that something like that described in the theory must be the case.

From this, Hunt develops a 'model' (not really a definition) to explain what we might mean by saying 'Theory X is likely to be true' – although because of point ii) above we cannot be certain that Theory X *is* true. The model presumes some theory (containing entities, attributes and structures) about the world. The theory has certain implications such as explanations or predictions which can be compared with the external world. These comparisons will result in successes or failures which reflect back on the theory. It is also recognized that the theory (or rather its implications) may have direct effects on the world through changing peoples' beliefs and behaviours. It is, then, the relative proportion of successes and failures which gives us cause to believe or not believe in the theory. The greater the proportion of successes, the more likely it is that something similar to the theory must actually exist.

Hunt then connect this idea to trust through the work of Harré (1986), who argued that, since we could never know for certain, in practice communities of scientists relied on trust. Rather than certain knowledge, they trade in trustworthy knowledge: that which they accept is genuinely believed, for good reasons, by their colleagues. For Hunt, this means that it is imperative that valid knowledge should have some grounding in the external world rather than being wholly internal to a particular discourse.

The argument that I wish to put forward is that both Meckler and Baillie's argument and Hunt's present a singular or monovalent view of truth and knowledge. That is, that there is only one type of truth – some weak form of correspondence between theories or beliefs and an external world; that there is only one way of evaluating truth; and that there is, therefore, only one form of knowledge, assuming that truth is an essential attribute of knowledge. In contrast, I shall suggest that there are several quite distinct forms of knowledge and correspondingly different ways in which truth claims may be redeemed.

8.3 Knowledge management

The discussion so far has been at the level of management knowledge in general, that is, the knowledge produced by management research, but to focus the debate I want to drop to one particular area of management: that of knowledge management. We are used to the idea of managing *information*, within information systems (IS) and increasingly on the Web, but it is also argued that within organizations people generate and use *knowledge*, as distinct from information, to preperly perform their work. This has led to the development of knowledge management systems as well as information management systems.

This is an interesting domain to consider for two reasons. First, as its object is in fact knowledge itself we might expect that it would be particularly concerned about clarifying and defining the nature of knowledge and, equally, truth. Second, knowledge management is not only interested in academically produced knowledge but also in the everyday, practical knowledge of managers and employees. This forces us to consider knowledge in a wider sense than just the output of academic research.

Within knowledge management (KM) it is conventional (Bell 1999; Boisot 1995; Davenport and Prusak 1998; Freeman 2001) to draw up a ladder from *data* to *information* to *knowledge* – what Tuomi (1999) calls the knowledge hierarchy. This is mirrored historically within IS in the move from *data* processing to *information* management to *knowledge* management. To give some examples, for Davenport and Prusak (1998) *data* are discrete facts about the world which in themselves are meaningless; *information* is data that has been processed or interpreted within a particular context to inform or reduce uncertainty; while *knowledge* is information that is even more valuable because of the addition of insight, experience, context or interpretation (Grover and Davenport 2001). Others who use the same basic model define knowledge in different ways. For example, knowledge is that which enables us to assign meaning to data (van der Spek and Spijkervet 1997); knowledge consists of truths, beliefs, concepts, judgements and expectations (Wiig 1993); or knowledge is tested, validated and codified information (Earl 1994).

Tuomi (1999) actually argues the case for a reversed hierarchy, namely that knowledge precedes information which in turn precedes data. On this view, knowledge becomes articulated within a verbal and textual context to form an information structure. This may be embodied as a document, a diagram, a data structure or information system. Once this has become totally defined the 'meaning' of the information is essentially fixed and this allows it to be populated or instantiated with items of data which would, by themselves, have no meaning at all. Put the other way round, data cannot exist without a pre-defined semantic and syntactic structure, which is information; and information is the articulation or explication of knowledge.

Other authors have developed more complex categorizations of knowledge (Marshall and Sapsed 2000). Millar *et al.* (1997) concentrate on what the knowledge is about and specify know-what, know-why, know-how, know-who, and experiential knowledge which can involve any of the others. Blackler (1995), drawing on Collins (1993), focuses on where the knowledge is situated and distinguishes between knowledge that is embrained (cognitive), embodied (perceptual), encultured (social), embedded (systematized), and encoded (formal or symbolic). This has been applied empirically by Thompson and Walsham

(2004). Other classifications have been suggested by, for example, Winter (1987), Fleck (1997) and Benson (Benson and Standing 2001). Many writers (e.g. Stenmark 2001; Tsoukas and Vladimirou 2001) refer to the distinction between tacit knowledge and focal knowledge originated by Polanyi (1958) and popularized by Nonaka and Takeuchi (1995).

However, as has been pointed out by many commentators (Swan and Scarbrough 2001), the nature of knowledge itself is highly debatable and several authors are critical of the whole emphasis on knowledge as some objective, commodifiable entity. Alvesson and Karreman (2001, p. 995) argue that knowledge 'is an ambiguous, unspecific and dynamic phenomenon, intrinsically related to meaning, understanding and process and therefore difficult to manage'. Marshall and Sapsed (2000, p. 12) emphasize the 'importance of considering knowledge not simply as a stable and unproblematic object that can be effectively decontextualized and freely circulated, but as a complex, dynamic, and situated series of processes'. In addition, they go on to argue that knowing is essentially active – to be able to act effectively within a social situation.

In practice, though, the overwhelming approach within knowledge management is to take a resolutely functionalist reading of knowledge, as Schultze and Leidner's (2002) research showed. They classified research articles on knowledge management between 1990 and 2000 into one of Deetz's (1996) four discourses of management – normative (functionalist), interpretive, dialogic (postmodern) or critical. Of the 75 papers, 71 per cent were classified as normative with a further 25 per cent being interpretive. Schultze and Stabell (2004) look at the contradictions involved in trying to manage tacit knowledge through Burrell and Morgan's four paradigms. There is not space in one chapter to provide a detailed critique of all these approaches to the definition of knowledge and information, so I shall make some general points that will illustrate what I see as their weaknesses.

With respect to the various versions of the knowledge hierarchy I would argue that they all suffer from inadequate and unclear conceptualizations of the nature of information and its possible relationships to knowledge. Mingers (1996b) carried out a thorough survey of existing theories of information, many of which were based in some way on Shannon and Weaver's (1949) information theory, including socially sophisticated models such as Mackay (1969) and Luhmann (1990b). These theories were evaluated in terms of four criteria: the generality of the theory; the pragmatic usefulness for information systems; the degree of integration with other disciplines; and lastly the correspondence with everyday usage. The approach that was judged most successful was that of Dretske (1981), and this formed the basis of a new theory of information and meaning (Mingers 1995a). This theory used Dretske's idea of knowledge and the flow of information but incorporated concepts from Habermas's (1978, 1984, 2003b) theory of communicative action and Maturana's (1995) cognitive theories embedded within a critical realist philosophy (Mingers 2004c).

With respect to Tuomi's reversed hierarchy there are aspects of this that are valuable. Clearly knowledge does structure that which can be information for us, and conditions the amount or extent of knowledge that is available from a particular source. For instance, as Polanyi (1958) has shown, gaining information from an X-ray requires considerable knowledge. Equally, data does rely on a pre-existing and consensual semantic and syntactic structure for it to be effective as data. However, I will argue that we need both hierarchies – data can carry information and, in certain circumstances, this information can then generate knowledge. At the same time a subject's knowledge alters the information they can receive, and allows them to access the information in the first place. We thus need more of an interactive view.

With regard more specifically to theories of knowledge, then, there are three general problems. The first is the very large number of papers that take a simplistic and unquestioning view of knowledge as an objective commodity. Second, those authors who do recognize different forms of knowledge point out particular and partial sets of distinctions based either on the object of knowledge, the form of knowledge, or the location of knowledge and do not thereby do justice to the richness of ways in which we talk of 'knowing'. Third, and particularly important for this chapter, almost none of the literature considers the relation of knowledge to truth.

8.4 Truth

One of the most traditional debates in philosophy has been that of epistemology – that is the study of knowledge (*episteme*) as opposed to mere belief or opinion (*doxa*). When are we entitled to say *I know* something rather than merely *I believe* it? We may all believe certain states of affairs to be the case, or that we know how to do certain things, but ultimately in order to be *knowledge* these beliefs must be testable or able to be validated in some way; that is, there must be grounds for them to be considered to be *true*.

It is interesting and perhaps indicative of the field that there is almost no discussion at all, within the knowledge management literature, of the problems of truth or warrantability. The assumption seems to be made that either knowledge is no different from any other cognitive category such as thought or belief, or that determining whether something is or is not knowledge is outside the scope of knowledge management. Even one of the founders of KM, Machlup (1980), went in this direction. Having produced an informed discussion of varieties of truth and truth-seeking, and discussed other aspects of the quality of knowledge such as beauty (aesthetics) and ethicality (axiology), he declared that such issues were largely irrelevant to his task of analysing the production and distribution of knowledge (p. 117).

Freeman (2001), whose paper entitled 'IS Knowledge: Foundations, Definitions and Applications' seems to promise some answers, defines knowledge loosely as 'information that has been validated and is thought to be true' with no consideration of what being true might mean. Von Krogh and Grand (2000) are concerned with organizational knowledge, and in particular the way in which new knowledge comes to be accepted or rejected by an organization especially where it contradicts the existing knowledge base.

Tell (2004) recognizes that knowledge requires some justification for it to be knowledge. He does not look to theories of truth for this; instead he looks at the contexts within which particular knowledge claims may be justified. He recognizes two dimensions to this: external vs. internal and procedural vs. performative. External justification relies on reference to some causal reality external to the knowing subject, while internal justification depends on the extent to which a belief is coherent with other beliefs and assumptions that already exist. With procedural justification, valid knowledge is generated by the extent to which a particular procedure or methodology is followed. Classically, scientific knowledge receives its justification from being the result of a scientific method. In contrast, knowledge justified by performance does not stem from following rules and procedures but from imagination, intuition, action and ultimately performative success. Tell than uses these distinctions to demarcate four forms of knowledge: objective, subjective, personal and organizational.

8.4.1 General theories of truth

I will now summarize the main theories of truth as found in philosophy before considering in more detail the theories of truth coming from critical realism and Habermas. Note that most theories concern the truth of propositions about states of affairs in the world.

The most common view, in Western philosophy anyway, is that knowledge is justified, true belief (JTB). This stems from Plato's *Theaetetus*, where Socrates argues that

> When, therefore, anyone forms the true opinion of anything without rational explanation, you may say that his mind is truly exercised, but has no knowledge; for he who cannot give and receive a reason for a thing, has no knowledge of that thing; but when he adds rational explanation, then, he is perfected in knowledge.

Socrates was sharp enough to point out later the self-referential difficulty of 'knowing' what is a rational explanation. These three conditions have been taken to be both necessary and sufficient for a proposition to count as knowledge. In other words, to validly assert 'I know that *p*' implies:

- You must sincerely believe that *p* is the case.
- You must have justifiable grounds, evidence, or explanation for *p*.
- *p* must, indeed, be true.

Although this sounds clear, there are in fact many problems with each condition as well as their conjunction. For instance, there is much debate about what would constitute proper justification for such a belief – empirical evidence, rational argument, personal experience, perception or what? How in any case can we determine if something is actually true? There are a whole range of theories of truth – correspondence, confirmation, coherence, or consensus, not to mention sceptics (e.g. Rorty 1989) who would deny the possibility of truth in the first place. Indeed we might say that the question of truth is actually the same question as that of knowledge, so defining knowledge in terms of truth makes little progress.

There is also the Gettier problem that provides cases where each of the conditions holds but we would still not wish to call it knowledge (Gettier 1963). For example, suppose you walk in to a room and think you see your friend John. You believe it to be John; you have grounds for believing it (he looks like John); but suppose you are mistaken and it is actually John's twin brother Mark. It would appear that the third condition is not met and you do not therefore 'know' that John is in the room even though you believe it to be the case. Suppose, however, that John *is* actually in the room but hidden around the corner so you do not see him. Now the third condition becomes true, even though you are not aware of it, and so you are entitled to say you know that John is in the room, even though you are actually mistaken in identifying John!

I shall briefly summarize the main philosophical theories of truth:

- Correspondence theories (Popper 1959; Russell 1912; Tarski 1944; Wittgenstein 1974) are the main and most obvious view of truth. They hold that truth (and falsity) is applied to propositions depending on whether the proposition corresponds to the way the world is. It thus applies to the relationship between a proposition and the states of affairs it describes. Problems with this view are: (i) In what sense can a linguistic statement be said to correspond to something quite different – an occurrence in the world? (ii) We cannot directly access the external world so we are only ever comparing experiences and statements with other experiences and statements, so that we can never actually determine if a proposition is, in fact, true. Most other theories stem from the problems in maintaining a correspondence theory.
- Coherence theories (Bradley 1914; Putnam 1981; Quine 1992) stress the extent to which a proposition is consistent with other beliefs, theories and evidence that we have. The more that it fits in with other well-attested ideas the more we should accept it as true. This approach avoids the need for a direct comparison with 'reality'. However, it is more concerned with the justification of beliefs rather than their absolute truth. From a Kuhnian (1970) perspective, fitting in with the current paradigm does not make the current paradigm correct. Quine held that coherent systems of beliefs were under-determined by empirical data and thus that no theory could ever ultimately be verified or falsified.
- Pragmatic theories (James 1976; Peirce 1878; Rorty 1982) hold that truth is best seen in terms of how useful or practical a theory is: the theory that best solves a problem is the best theory. A version of this is instrumentalism, which holds that a theory is simply an instrument for making predictions, and has no necessary connection to truth at all. This also leads into

consensus theories. An obvious argument against this view is that a true theory is likely to be most useful and powerful[2] and therefore should be an important component of a useful theory.

- Consensus or discursive theories (Habermas 1978) accept that truth is that which results from a process of enquiry resulting in a consensus among those most fully informed – in the case of science, scientists. At one level, we can see that this must be the case if we accept with critical realism the impossibility of proving correspondence truth. But, often, today's accepted truth is tomorrow's discarded theory and so this does not guarantee truth. See the discussion below about Habermas's more recent views.
- Redundancy and deflationary theories (Frege 1952; Horwich 1991; Ramsey 1927) argue that the whole concept of truth is actually redundant. If we say 'it is true that snow is white' we are saying no more than that 'snow is white'; the two propositions will always have the same truth values and are therefore equivalent. This seems to me largely a linguistic move as it does not touch upon the question of how we might know or believe that a proposition is actually the case.
- Performative theories (Strawson 1950) also deal with the linguistic use of the term. The suggestion here is that by saying '*p* is true' we are not so much commenting on the truth of the proposition as such but on our willingness or intention of accepting it as true and commending it to someone else. Again, this just seems to ignore large areas of the question of truth.

8.4.2 Critical realism and truth

Turning now to critical realism, what view of truth does it espouse? First, how does critical realism (CR) relate to Hunt's scientific realism? It is simplest to describe it as a version of scientific realism. Certainly Bhaskar would accept Hunt's four propositions, and has in fact written a book called *Scientific Realism and Human Emancipation* (Bhaskar 1986). The first thing to say about truth is that the whole approach is fallibilist, as in Hunt's principle ii). That is, since it accepts epistemic relativity, the view that all knowledge is ultimately historically and locally situated, it has to accept that theories can never be proved or known certainly to be true. Thus, if provable truth were to be made a necessary criterion for knowledge there could be no knowledge within critical realism.

Bhaskar does discuss the notion of truth and comes up with a multivalent view involving four components or dimensions (Bhaskar 1994, p. 62) that could apply to a judgment about the truth or falsity of something.

1. *Normative-fiduciary:* Truth as being that which is believed from a trustworthy source – 'trust me; I believe it; act on it.' This sense would typically occur within a communication where the speaker states a proposition and the listener accepts their sincerity. This is clearly related to Hunt's argument about trust and also stems ultimately from Harré.[3]

2 *Adequating:* Based on evidence and justification rather than mere belief – 'there's sound evidence for this.' This goes beyond just the speaker's belief to warranted assertability but can still, of course, be false.
3 *Referential-expressive:* Corresponding to or at least being adequate to some intransitive object of knowledge. Whereas the first two dimensions are clearly in the transitive domain and strongly tied to language, this aspect moves beyond to posit some sort of relation between language and a referent. It moves towards a weak correspondence theory.
4 *Ontological and alethic:* This final level is the most controversial (Groff 2000), as it moves truth entirely into the intransitive domain: the truth of things in themselves, and their generative causes, rather than the truth of propositions. It is no longer tied to language although it may be expressed in language.

Several comments need to be made here. First, the first three dimensions are relatively unproblematic and quite similar to the JTB formula, although set within a communicative context. 'This proposition is believable' (B); 'don't just listen to me, there is some evidence for it' (J); and 'it fits the facts' (T); none of these in themselves or, indeed, together *guarantees* that it is true.

Second, Bhaskar sees them as ordered or progressive. Thus, the weakest form of truth is simply to have to believe someone with no further justification. Better is to have some sort of warranted assertability, some evidence justifying the claim, although what the evidence is and how strong it is are debatable points. Better still, there should be some theory, description, or model that can be related to real-world structures. This obviously moves in the direction of some sort of correspondence theory of truth. Critical realism does tend towards this view while accepting inevitable limitations on it (Sayer 2000).

Third, the ontological/alethic aspect marks a major shift, as it no longer concerns propositions at all. It is not predicated on a proposition but is said to be a characteristic of the 'real' nature and causes of things in themselves: 'truth as alethic, i.e. the truth of or reason for things, people and phenomena generally (including in science most importantly causal structures and generative mechanisms), not propositions' (Bhaskar 1994, p. 64).

8.4.3 Habermas's theory of truth

We can now move to consider Habermas's theories of knowledge and truth. His early work is known as the theory of knowledge-constitutive interests (Habermas 1978). This suggested that humans, as a species, had needs for, or interest in, three particular forms of knowledge. The *technical* interest in moulding nature led to the empirical and physical sciences. For Habermas these were underpinned by a pragmatist philosophy of science (inspired by Peirce) and a consensus theory of truth. The *practical* interest in communication and mutual understanding led to the historical and interpretive sciences, underpinned by a hermeneutic criterion of understanding. And the *emancipatory* interest in self-development and authenticity led to critical science, which identified repressions and distortions in knowledge and in society. Its criterion of success was the development of insight and self-expression free from constraint.

This theory of transcendental interests was the subject of much criticism (see Mingers 1997c for a review), and Habermas later transmuted it into the theory of communicative action (Habermas 1984; Habermas 1987). Utterances and, I would argue, actions as well raise certain validity claims that must, if challenged, be justified. These claims are *comprehensibility*, *truth*, *rightness* and *truthfulness* (*sincerity*). This is premised on the argument that utterances stand in relation the three different 'worlds' – the objective or material world that consists of all actual or possible states of affairs; the social or normative world that consists of accepted and legitimate norms of behaviour; and the subjective or personal world that consists of individuals' experiences and feelings.

When such a claim is challenged, the process of justifying the claim must always be discursive or dialogical; that is, there should ideally be a process of open debate unfettered by issues of power, resources, access and so on until agreement is reached by the 'unforced force of the better argument' (Habermas 1974, p. 240; Habermas 2003b, p. 37), what Habermas calls the 'ideal speech situation' (see Chapter 10). Thus, Habermas held a consensus or discursive view of truth both in the moral or normative domain of what ought we to do, as well as in the material domain of external reality. To say of a proposition, 'it is true' is the same as saying of an action, 'it is right', namely *ideal, warranted assertability*. This links up to the realism debate discussed earlier, since a constructionist such as Gioia would clearly be committed to a consensus theory of truth without recourse to an external world.

However, more recently Habermas (2003b) has returned to the issue of truth and now rejects his discursive theory for propositions about the material world in favour of one with an irreducible ontological component. In essence, Habermas now maintains that there is a substantive difference between the moral domain of normative validity, which can only ever be established through discussion and debate within an ideal speech situation, and the domain of propositional truth where properly arrived at and justified agreement may still be proven wrong by later events. 'I have given up an epistemic [based only on reason and discussion] conception of truth and have sought to distinguish more clearly between the truth of a proposition and its rational assertability (even under approximately ideal conditions).' (Habermas 2003b, p. 8.) Habermas now accepts the basic realist view that there is a world independent of humans, that we all experience the same world, and that this places constraints upon us, while still accepting that our access to this world is inevitably conditioned or filtered through our concepts and language.

This, of course, leads to the age-old dilemma of trying to discover some external standpoint, outside of language and cognition, from which to judge the truth of one's propositions. The idea of ideal rational discourse is not wholly wrong, but it is insufficient for the task (p. 252). While it is necessary that we come to believe or accept the truth of propositions through a thorough process of rational discourse, that we do so is not sufficient to guarantee their truth. Even the most strongly held and well-justified views may turn out to be false. 'These objections have prompted me to revise the discursive conception of rational acceptability by relating it to a pragmatically conceived, nonepistemic concept of truth, but without thereby assimilating "truth" to "ideal assertability".' (Habermas 2003b, p. 38.) The basic outline of this nonepistemic concept of truth has a very Popperian ring to it. If we begin with our everyday purposeful activities within the lifeworld, we can see that our perceptual and conceptual apparatus unavoidably shapes our access to reality – we never meet it naked – but at the same time our interactions, and particularly our failures, lead us to revise our conceptual structure. In the lifeworld, while engaged in action, we presume and do not question the truths of the propositions we operate under. Only when these break down do we move from action to discourse and offer our beliefs up for debate and justification. Once we have become convinced of the truth of a proposition through the process of rational discourse we can then move back and adopt it within the sphere of engaged action. It is important in this process that the reasons we adduce for coming to believe a proposition are actually related to the experiences that have led us to question and debate. Within the true, justified belief definition of knowledge, the justification must stem from the actual experiential learning that has occurred rather than being ad hoc or coincidental as in the Gettier example above.

Habermas's move away from an epistemic (discursive) conception of truth is actually towards an ontological one. When we make what we take to be true assertions we are expressing beliefs that certain states of affairs do actually exist, and that these in turn refer to entities or relations that also exist. This establishes a relation between truth and reference: between the truth of statements and aspects of an objective world. This is so even between different linguistic communities (spatial or temporal) where the same referents, the same objects of discourse, may well go under different descriptions. 'The experience of "coping" accounts for two determinations of "objectivity": the fact that the way the world is is not up to us; and the fact that it is the same for all of us.' (Habermas 2003b, p. 254.) This does not of course guarantee that the 'knowledge' is true – Habermas is fallibilist in the same way that Bhaskar is: 'Insofar as knowledge is justified based on a learning process that overcomes previous errors but does not protect from future ones, any current state of knowledge remains relative to the best possible epistemic situation at the time.' (Habermas 2003b, p. 41.) Habermas's move is certainly welcome from a realist position. One criticism was always that his view of natural science was overly pragmatic or even instrumental. He tended to call it 'empirical-analytic' and this, combined with the consensus theory of truth, lost touch with a realist view of ontology. It also meant that he was essentially anti-naturalist, seeing a radical disjunction between natural science and social science. This shift to some extent addresses both issues: accepting a causally constraining reality as discussed above; and accepting a 'weak naturalism' (Habermas 2003b, p. 22) that there is an underlying evolutionary continuity between the objective world and the lifeworld, between nature and culture.

However, I would argue that Habermas does not go far enough in this direction, and more specifically remains too strongly wedded to the idea that validity claims, including those of (nonepistemic) truth, are validated linguistically. In the model described above, problems and failures in the world of action lead to a switch to the world of discourse wherein questions of truth are decided through debate. Now while I accept that humans do always interact within language, that is not to say that all activity is linguistic. Within the realm of epistemological knowledge (i.e. science), experimental activity is clearly the cornerstone of progress. With performative knowledge, the measure is successful performance whether it is a motor skill such as riding a horse or a social skill such as conducting a meeting. And with experiential knowledge, claims to have had a particular experience can be investigated forensically, i.e. through some form of 'detective' work. Thus, the results of activity and action will inform the linguistic debates.[4]

8.4.4 Summary

As can be seen, truth is a highly complex and debatable concept. I would like to pull out the following general conclusions in leading on to consider the relations between knowledge and truth.

The underlying conception of truth, supported by Hunt, Bhaskar and Habermas, is a limited form of correspondence theory. As realists, we accept the existence of an independent or intransitive domain of objects of knowledge that have causal effects and thereby confirm or disconfirm our knowledge. We also have to accept, however, that we can never have pure unmediated access to this domain and thus that our knowledge is always provisional and subject to change.

This places the emphasis on the degree of warrantability or justification that there is for something claiming to be knowledge. Is it a matter of believing a trustworthy source? Seeking supporting evidence? Witnessing a demonstration? Or conducting extensive scientific research?

Here, some of the other theories of truth come into play. As Habermas emphasizes, all truth claims are ultimately validated discursively through discussion and debate. Even when the intransitive world appears to refute some theory, say through failed experiments, it is the community of scientists who decide why the experiment is failing and at what point it becomes conclusive (Collins 1985). Thus, there is always an element of consensus about truth claims.

Another form of support is the extent to which a theory is consistent with other well-attested knowledge – i.e. its coherence. But of course we have to recognize that innovations often contradict the perceived wisdom. Success in practice (i.e. pragmatism) also provides support for a theory, although while a true theory should be successful it does not follow that a successful theory is true.

8.5 Forms of knowledge and truth

Much of the philosophical discussion of knowledge (an obvious exception being Ryle 1963) confines itself to propositional knowledge of a scientific nature.

However, as we have seen, knowledge management is concerned with knowledge in a wider, everyday sense. As one of the founders of KM said,

> Most philosophers confine their discourses on knowledge to verbal propositions;... In this book I use the word knowledge in a much wider sense, because a narrow (or 'strong') sense of knowing would restrict it meaning unnecessarily. If Tom, Dick and Harry say that they know somebody or something ... it is not reasonable to insist on so restrictive a definition of knowledge that 90 per cent of all that Tom, Dick and Harry claim to know is 'really' not knowledge.
>
> (Machlup 1980, p. 97)

So I am concerned with the ways in which the word 'knowledge' or, more actively, 'to know' are used in everyday speech: 'I know her well'; 'I know how to ride a bike'; 'I know there's a train at 3.00'; I know I left my key there'; 'I know the feeling'; 'I know what black holes are'; 'I know how to make a presentation'; 'I know how the system works'; 'I know linear algebra', 'I know how to speak Italian'. This leads us to consider whether there may be other important forms of knowledge with different characteristics and means of validation.

Generally, I will be talking about knowledge in the personal sense (Polanyi 1958), that is, in terms of an individual and what they know, either consciously or unconsciously. Knowledge also exists in an extra-personal sense as embodied in books, papers, films, organizational practices and procedures, the internet etc. – World 3 in Popper's sense (Popper 1972) – and indeed much of the literature in knowledge management concerns precisely the interaction between the two (Tsoukas and Vladimirou 2001). This can be viewed as the dualism between action and structure in Giddens' (1984) structuration theory terms. In this analysis I will concentrate on the 'action' side of the dualism, that is, the way individuals come to 'know' and then act in relation to this knowledge. In taking this view I am using Giddens' 'methodological epoché' (Giddens 1979, p. 80) to bracket one side of the duality, the way that structural (e.g. organizational) knowledge is generated and reproduced, in favour of the other side: 'the individual ability to draw distinctions within a collective domain of action, based on an appreciation of context or theory, or both' (Tsoukas and Vladimirou 2001, p. 979).

To discover as many different senses of the term 'to know' in everyday use I reviewed a wide range of dictionaries. This identified thirteen[5] distinguishable uses, as shown in Table 8.1. Clearly there are certain families of resemblances here (Wittgenstein 1958) and we will classify them in a later section, but to begin with we can identify certain dimensions that all usages of 'to know' have in common.

First, any form of knowledge must be knowledge *of something* (Bhaskar 1978). There must always be an *object of knowledge* although by no means necessarily a material or physical object. In the above examples, such objects include states of affairs, people, skills, values, feelings and emotions, social

Table 8.1 Senses of the term 'to know'

No.	*Dictionary definition*	*Example*
1	To perceive directly, to have cognition of	I know it's raining
11	To have full information of	I know everything there is to know about widgets
12	To know things from information	I know there's a train at 5.32
4	To recognize as the same or as familiar	I know that voice
5	To be acquainted with (people) (organization)	I know your mother; I know your school
8	To have experience of	I've known the cares of office
13	To be acquainted with emotions and situations	I know the feeling; I know how you feel; I know how stressful an exam is
7	To have practical understanding of; to have a skill	I know how to use Excel; I know how to play the piano
9	To have fixed in the mind; to learn	I know my French verbs
2	To have understanding of	I know how a diesel engine works
3	To recognize the nature of	I know that lump's benign
6	To be aware of the truth of; to be certain of	I know she's lying
10	To be able to distinguish	I know right from wrong

practices, organizations, and complex physical entities. Nevertheless, there must be some object of knowledge and this connects immediately with critical realism. Knowledge itself, especially as it is an individual person's knowledge, is always in the transitive dimension but the objects of knowledge, even where they are concepts or ideas, are intransitive – objects available for investigation or debate. This transitive/intransitive distinction is not fixed as in the internal/external world, but is always relative to the event and context. Thus even an utterance, which is initially in the transitive domain, becomes intransitive after it has been uttered as it can then become an object of knowledge: 'Did you say that?'; 'What did you mean?'

Second, there must always be a *source of knowledge* – knowledge must come from somewhere, generally some aspect of a person's experience. Some possible sources of knowledge are direct perception; a message or communication; reading a timetable or book; learning, as in a language; practice, as in playing the piano; or plain experience over time. It is here that the most direct connections with information and meaning come – information can be a source of knowledge, and existing knowledge shapes the information that is available from a source, as the example of reading an X-ray shows.

Related to this is the third dimension – the way the knowledge is stored or represented, particularly in terms of the degree of tacitness/explicitness. Some knowledge will be entirely conscious and explicit – we know that we know it and we can express it clearly. Some knowledge will have a degree of tacitness (Polanyi 1958) – we have the knowledge but are not necessarily fully conscious of it, or fully able to articulate it. For instance, we can speak a language without knowing the rules that govern it; or we can use a carpenter's plane and know when the blade needs changing. Finally, much of our knowledge, especially at a perceptual/motor level but also at higher levels as well (Mingers 2001b), is *embodied* at a pre-conscious level. It governs or shapes what we can be conscious of (Merleau-Ponty 1962; Varela *et al.* 1991).

Fourth, as we have already discussed, one very important facet of knowledge is its *truth* or warrantability. This is supposedly what distinguishes knowledge from simply belief or opinion. However, the nature of truth is a very complex question and differs between different forms of knowledge so I shall discuss this in the next section.

This has led me to distinguish four generic forms of knowledge which differ in terms of the above four characteristics and yet account for the thirteen usages identified in Table 8.1. These are summarized in Table 8.2.

8.5.1 Everyday propositional knowledge

This form of knowledge is our everyday, common-sense, relatively direct awareness of the world around us. To know in this sense is to know *that* – to be aware of or to be cognizant of states of affairs. It is to know that it is raining, that there is someone at the door, that there is a train at 12.15, that there are 35 widgets in stock, or that the petrol tank is half-full. I call it propositional knowledge, in comparison with the other forms, because it is generally explicit and conscious, and can be represented in the form of propositional statements: 'I know that *x* is or was the case' (Klein 1971).

We gain propositional knowledge from several sources. This first is our direct perception of the world, through any of the senses. In philosophy, this kind of direct knowledge of things is called *de re* as opposed to that which we are told about – *de dicto*. In fact, Dretske, whose work we drew on earlier, actually restricts his theory of knowledge to only this kind of direct perceptual knowledge generated by the receipt of signs carrying information. But I shall include more generally knowledge of states of affairs that we are told about through a linguistic (or indeed non-verbal) communication, and knowledge we get through books, papers, timetables and so on.

In terms of its warrantability, propositional knowledge is *referential-expressive* in Bhaskar's terms and concerns the validity claim of *truth* in Habermas's communication theory or non-epistemic truth in his later work. Here we can go beyond belief and even justification towards confirming a relation between the proposition and the intransitive world to which it refers.

Indeed, if we follow Dretske (1981) and Mingers (1995a) we can see a direct *causal* relation between information and the propositional knowledge that it creates. Dretske argues that the meaning which is generated from the information we receive leads us to have certain beliefs about the world. Now, beliefs as such are not identical to knowledge, as is shown by the justified, true belief formula mentioned above. We may genuinely believe something but it may in fact be false even though there is justification for it, e.g. the belief that the sun orbits the earth. Or, we may believe something that is in fact true but for the

Table 8.2 Forms of knowledge and truth

Type of knowledge	*Object of knowledge*	*Source of knowledge*	*Form of representation*	*Criteria for validity*
Propositional – I know it's raining; I know there's a train at 3.00; I know there's someone at the door	States of affairs in the physical and social world. *To know that x*[1]	Direct perception, receipt of information, communications, the media	Generally explicit and propositional although some may be tacit	*(Ontological) truth*[1] **Referential-expressive**[2]
Experiential – I know her well; I know the feeling; I know I left my key there; I know how the system works	People, places, events we know through personal experience. *To know x*[1]	Personal experience	Memories, some aspects of which may be tacit and embodied	*Sincerity*[1] **Normative-fiduciary Adequating**[2]
Performative – I know how to ride; I know how to read an X-ray; I know how to make a presentation	Skills, abilities and competences. *To know how to do x*[1]	Personal experience, learning, training	Embodied	*Competence, (epistemic) rightness*[1] **Alethic**[2]
Epistemological – I know what black holes are; I know linear algebra	Reasons for the (non-) occurrence of things and events. *To know why x*[1]	Formal methods of discovery, e.g. in science	Explicit, discursive, 'objective', open to debate	*Truth, rightness, sincerity*[1] **Ontological, alethic**[2]

Notes
1 Habermas's validity claims.
2 Bhaskar's four dimensions.

wrong reasons, e.g. we may think it is 10.00 am (and it is) because the clock says so, although the clock has in fact stopped. In neither case would we be entitled to say 'we know that…'. However, beliefs caused by information must be true since for Dretske information must be true to be information. Thus if the clock were working correctly we could say we know it is 10.00 because our belief is caused by (true) information.

Even so, we cannot finally prove our knowledge is true for we might be mistaken either in our interpretation of the sign (misreading the clock), or in believing it was (true) information when in fact it was not (the clock was stopped).

8.5.2 Experiential knowledge

We talk about knowing in this sense when we are referring to our own individual prior experience, particularly of people, places, events or feelings. To know in this sense is to be acquainted with or to be familiar with. Thus, I know Mary Scott, I know Birmingham, I know 'that feeling', I don't know your school, or I know how bad toothache can be.

Knowing in this sense is a statement about the experience that someone has had, or not had, in the past. The depth of knowledge concerned is very variable – in saying 'I know Mary Scott' I might just mean I know who she is, or I might mean that I know her very well. This form of knowledge is not primarily propositional. We can always make a propositional statement about it – 'It is true that I know Mary Scott' – but this is a second-level statement, the object of which is our first-level experiential knowledge. We do not say 'I know *that* Mary Scott.'

Knowledge in this sense can be much richer and deeper than simple propositional knowledge. To know someone is not simply to know that they exist, it is to have a complex set of understandings, experiences, feelings and beliefs about that person. Much of this may be tacit and difficult to express explicitly. It is also deeply personal or subjective, since my experience of a person or place may be very different from someone else's.

The validity of this form of knowledge must ultimately come down to a matter of Habermas's *truthfulness* or *sincerity* (*normative-fiduciary* in Bhaskar's terms) since it concerns a particular person's experiences or feelings. Of course, one does not just have to accept a person's discursive justification; one might try to discover or provide some sort of evidence or justification as well, which could include documentary evidence – letters, photos, transcripts, etc., or corroboration from other people.

8.5.3 Performative knowledge

Performative knowledge involves having some skill or competence in order to be able to do something – it is to know *how* rather than to know or to know *that* (Polanyi 1958; Ryle 1963). I include in this category much more than simple physical skills. So, we can talk of knowing how to ride a bike; knowing how to play the piano; knowing how to speak a language; knowing how to 'play the game' as in office politics or a sport; knowing how to parent; or knowing how to cook.

What distinguishes this type of knowledge is that it goes beyond simple experience of something to involve particular skills and abilities that have to be learnt over a period of time. It generally involves explicit training in order to develop the necessary skills. I call it performative because it usually involves some kind of physical motor skills, some kind of performance – it goes beyond knowledge in a purely conceptual sense. For example, one could know plenty of the theory of music without being able to play the piano, and in its turn playing the piano does not mean that you can play the violin. Each skill has to be learnt over time and through practice – it is inscriptive rather than intellective (Hayles 1992). It may also involve considerable amounts of time – Gladwell (2008) estimates that exceptionally good performers, e.g. concert pianists or sports stars, have probably spent 10,000 hours practising and playing to be that good.

This in turn means that performative knowledge is inherently embodied (Mingers 2001b; Varela *et al.* 1991) – that is it exists as dispositions or connective states of the body and nervous system itself and may well be pre-conscious. Even higher-level skills such as language (Lakoff and Johnson 1987; Merleau-Ponty 1962; Merleau-Ponty 1969) or cognitive/mathematical activities such as navigation (Hutchins 1995) have significant bodily aspects. I once observed, at an airport, an English girl talking to her English friend. Their conversation was typically quiet and low-key. She then struck up a conversation with an Italian woman and it turned out she was herself half Italian and could speak Italian. Her whole manner and disposition changed instantly, becoming louder, more emotional, and much more animated as she unconsciously switched from *being* English to *being* Italian.

Generally, experiential knowledge is evaluated in terms of practical success or failure rather than truth. Can one actually ride the bike, play the piano or converse in Italian? Of course, there will be degrees of ability in many of these activities. Dreyfus (1992) presents a useful analysis from a phenomenological viewpoint of the development of skills from novice to expert. In some ways, this is actually quite close to Bhaskar's concept of alethic truth that I critiqued above. To demonstrate that one is a pianist by actually performing validates itself without need of propositions or assertions. We can also bring in here Habermas's validity claim of comprehensibility. Before a speech act or indeed any other social action can be judged it must be understood: that is, it must be performed in a competent manner. Habermas draws on Chomsky's (1957) notion of a competent speaker of a language (Habermas 1979, p. 29) but this can be enlarged to cover all the aspects of performative knowledge.

8.5.4 Epistemological knowledge

By epistemological knowledge, I am signalling a move away from the everyday *knowing that* things are the case towards deeper understandings of *why* things are as they are. It is to know *why*, to be knowledgeable about, to know the truth of, to be certain of, or to understand. It can be seen as related to or a development of everyday propositional knowledge and I would include within this category what we call scientific knowledge – very much the subject of critical realism. I have called it epistemological knowledge to indicate that it is the most self-conscious about its validity and, more than the other forms of knowledge, is centrally characterized by its concern for truth. It should not be confused with Habermas's epistemic (discursive) approach.

This form of knowledge goes beneath the surface of what appears to be the case, the domain of the empirical, to be able to account for the empirical in terms of underlying reasons or causes. I would not want this to be seen in terms of some simple-minded, linear model of cause and effect. Examples here are: to know how a diesel engine works, to know why inflation is falling, to know the difference between right and wrong, or to know 'What Freud Really Said', to quote a well-known book.

This type of knowledge is in some ways the obverse of performative knowledge, as it is almost entirely explicit and discursive and is judged in terms of its correctness rather than its success. It can be knowledge of an everyday kind – knowing how something works – but in the main it refers to scholarly knowledge that is generated according to well-defined procedures or methodologies. However, I do not include only knowledge of material things. Of equal importance (Habermas 1984; Habermas 1990) is knowledge of the social world and the personal world. In the social world we are interested in explaining why certain norms or patterns of behaviour exist and are maintained, and perhaps why others are not. In the personal world one wants to gain both valid interpretations of others and undistorted understanding of one's self (Sayer 2000).

I should like to end with one final comment. The chapter has been concerned with analysing several different forms or types of knowledge, but of course in real-world situations and activities these different types will typically be involved together and will interact with each other. To take just one example, suppose you are chairing a meeting. This will draw on propositional knowledge about particular facts and states of affairs; experiential knowledge of people, events, and practices; performative knowledge, perhaps of body language and physical gestures; and epistemological knowledge, perhaps of economics or a particular industrial process.

8.6 Conclusions

The first conclusion is to recognize the multidimensional nature of what we can know. Most knowledge management literature implicitly assumes that knowledge is an integral, easily definable commodity that can be extracted, stored and transmitted relatively easily. The literature that does not make this assumption either presumes knowledge to be some form of processed information; or categorizes it on a single dimension such as tacit/explicit; or argues that it is too complex to manage at all. In contrast, this chapter has proposed a polyvalent view of knowledge that recognizes four distinctively different forms of knowledge – propositional, experiential, performative and epistemological – based on several different dimensions. It is argued that this typology does justice to the rich and varied ways in which people may be said 'to know' something.

The second conclusion is to point out the intimate connection between knowledge and truth, which is rarely discussed in the KM literature. Knowledge, to be knowledge rather than simply opinion, raises claims as to its truth or validity. Truth, too, turns out to be a complex concept and within the chapter it has been explored from a critical realist perspective. This perspective grounds its concept of truth in terms of correspondence to an external, independent reality but recognizes that epistemologically knowledge is always provisional and relative. If truth can never be known with certainty then great attention must be paid to questions of justification and warrantability. What would lead us to accept a knowledge claim – accepting the trustworthiness of the source; witnessing an event; gathering evidence; or its consistency with our other beliefs?

This leads to the view that the different forms of knowledge imply different forms of truth or, rather, different ways of justifying their claim to truth. Propositional knowledge of day-to-day states of affairs can be directly justified in terms of the (true) information that generates it. Performative knowledge can be justified by a successful performance. Experiential knowledge can be justified through the sincerity of the claimant or the discovery of adequate evidence. Epistemological knowledge brings in the full force of science, whether it be natural or social.

I would like to make it clear that although this chapter has concentrated on the subjective aspects of knowledge – the knowing subject – and has primarily developed somewhat static categorizations, I see this as only part of a much broader domain that is both processual and social. In terms of process, events in the world carry information and lead to experiences that generate meaning, ideas and knowledge for individuals. At the same time, as Tuomi (1999) indicated, our knowledge, and more generally our cognitive structure, conditions both how we experience events and what information is available to us from them. This dynamic interactive process involves the material world but even more significantly the social world. As individuals, we exist in multiple social networks or forms of life (Wittgenstein 1958) and much of our everyday knowledge is actually intersubjectively shared knowledge about acting effectively within these social systems.

Notes

1 'Critical realism' as mentioned by Hunt, does not refer to Bhaskar's critical realism but is quite similar in character.

2 Although postmodernists argue that it is the theory that is deemed most powerful that is accepted as true.

3 Bhaskar was actually a student of Harré's and there are still clear resonances of this.

4 These different varieties of knowledge will be discussed later.

5 It is coincidental that Machlup (1980, p. 47) also identifies 13 elements of knowing – I had not read his book at the time. As one might expect, the two lists are similar but not identical.

Part IV

Practical issues

9 Methodology

What *might* we do?

9.1 Introduction

In the last few chapters we have discussed the two major questions of the philosophy of science – what entities do we take to exist that could be possible objects of knowledge and what might be their properties (ontology)? And what in fact do we mean by knowledge and are there limits on the knowledge that we can produce (epistemology)? This has, so far, been at quite an abstract level, but in this and the next chapter I want to move down to a more practical level and consider first research methods – the tools and approaches that we might use to generate knowledge – and second ethics – issues about what we *ought* to do as opposed to simply what we can do.

Research methods are specific techniques for generating different kinds of data, be it quantitative or qualitative (Bryman 2001; Easterby Smith *et al.* 2008; Kelemen and Rumens 2008). The term methodology has a range of different usages. At its most general, *methodology* is the study of methods as found in a methodology course or a chapter in a textbook. As such, it will cover a range of methods and how and when to use them. Most specifically, the term is used to describe the actual methods used in a particular research study – a PhD thesis will usually have a methodology chapter. There is, finally, an intermediate usage when a particular set of methods has become so well established that it can be called *a* methodology. For example, constructing, administering and analysing a questionnaire or the use of grounded theory could be said to be methodologies. Equally, within systems thinking, soft systems methodology (SSM) is a specific set of steps for taking action in problematic organizational situations.

Traditionally, there has been a tendency to link a particular set of research methods to a particular philosophical paradigm. Thus the empiricist or positivist paradigm, in both natural and social science, assumed the use of quantitative methods and statistical data analysis; the interpretive or constructivist paradigm in social science assumed qualitative methods such as interviews, hermeneutics, discourse analysis or participant observation. Critical paradigms used either type of method but adopted particular positions on issues such as the role of values in research and the oppressive nature of society. The rise of postmodernism in the 1980s and 90s led to skepticism about all forms of research methods and, indeed, about knowledge and truth itself. This led to a very divisive period in which the various paradigms fought against one another (the 'paradigm wars'), and all fought against postmodernism.

It was in reaction to this fragmentation, and in particular the skepticism from all sides concerning ontology and realism, that critical realism developed, and is, in many ways, the main reason that it has become so popular. In this chapter, we will begin by briefly reviewing the traditional research approaches – quantitative *or* qualitative methods – and particularly highlighting their limitations from a critical (realist) position. This will lead into a discussion of the more modern view that in fact research methods from different paradigms should be combined together in a research study – known as mixed methods research or multimethodology. Finally, we will discuss the specific contribution that critical realism brings, particularly to social science research.

9.2 Quantitative methods in systems research

By quantitative methods I mean any research methods that generate data that is in a measurable form and which is thus amenable to some type of statistical or mathematical analysis. Such methods generally rely on some kind of underlying model, be it deterministic, stochastic or a simulation. From an empiricist or positivist viewpoint, this is really the only valid form of scientific research; it dominated social science, particularly management and management systems, for many years. As we have seen in earlier chapters, alternatives have arisen, especially from an interpretive or constructivist paradigm, which have been highly critical of quantitative research, leading to hugely divisive splits within the discipline (Mingers 2004b). However, this often leads to a strongly anti-realist position that tends to deny the existence of any forms of external social structures (e.g. King 1999a).

This section puts forward a critique of traditional mathematical/statistical modelling and analysis but from a philosophical paradigm other than interpretivism – that of critical realism (CR). There are many different forms of quantitative models, so to provide a sharp focus for the critique, I shall take just one form of modelling as an exemplar – statistical forecasting, in particular multiple regression as used extensively within econometrics, information systems and management systems generally. The reasons for this choice are that that multiple regression is an extremely common technique taught and used across the management spectrum; and that it embodies many of the characteristics of statistical modelling in general, so much of the evaluation would apply to other types of statistical analysis (and even non-statistical modelling).

9.2.1 Prediction or explanation?

A good generic definition of a model is that given by Pidd (1996: 15): 'a model is an external and explicit representation of part of reality as seen by the people who wish to use that model to understand, to change, to manage and to control that part of reality'. While this certainly captures many elements of modelling it is, in a sense, too general – it obscures significantly different approaches to models and modelling. These can be seen more clearly in Mitchell (1993: 113), who distinguishes broadly two types: i) the model simply as a device for predicting outputs from inputs, there being no need for the model to *represent* the real system in any particular form; ii) the model as a statement of beliefs held about the world by some of those involved. Such a model will generally involve some sort of knowledge or at least beliefs about significant aspects of the real system.

It is clear from a CR perspective that the role of modelling should definitely be that of explanation and understanding rather than prediction. The argument can be made from two directions. On the one hand the inevitable limitations of data-only models, as in econometrics (Lawson 1997), means that their predictive accuracy will always be very poor in anything other than a very well defined and largely closed (perhaps mechanical or physical) system. More positively, trying to gain some understanding of the underlying causal mechanisms, no matter how difficult this is, will potentially provide a much richer understanding of the situation; the ability to consider possible changes that are outside of available or actual conditions; and ideas or theories that are potentially transferable to other situations, or that can be built on cumulatively. This use of modelling fits in well with CR retroductive methodology. The hypothesized mechanisms can be explicitly represented via models, such as those of system dynamics, discrete event simulation or even soft systems methodology (SSM), and the consequences of their operation can be studied.

9.2.2 Statistical modelling

Statistical analysis is a form of modelling that explicitly recognizes the existence of uncertainty in a set of data. It is conventionally seen as having two possible roles – descriptive and inferential. Descriptive statistics is simply concerned with summarizing the main characteristics of a dataset, particularly highlighting any patterns (and anomalies) that may not be immediately obvious. Inferential statistics goes beyond the data as given, recognizing that it is likely to be only a sample of all possible values (the population), to draw inferences from the sample to its underlying population.

From a CR perspective, descriptive statistics is unobjectionable and, indeed, very useful. If patterns exist within some set of observations (be they quantitative or qualitative) then there must be some underlying structures, mechanisms, or constraints generating them and this may prove a good starting point for a CR investigation. We must be aware, of course, that the patterns in the data cannot be assumed to simply reflect an underlying reality. Critical realism recognizes that the *process* of observation or, as we would call it, data *production* inevitably imposes itself on the results. This is especially so, we would argue (Mingers 1989b), within management (as opposed to a pure research situation), where much of the data we work with is produced on a routine basis within an organization, often for purposes different from the actual study, and with highly variable (and uncertain) levels of quality.

At first sight, inferential statistics would also appear to be compatible with critical realism in moving from actual observations to something underlying them, even if it is populations rather than mechanisms or structures. However, when we look at what is actually meant by this within statistics we find a very impoverished and empiricist viewpoint. A standard definition is 'inferential statistics consists of methods that use sample results to help make decisions or predictions about a population' (Mann 2004, p. 3). The 'results' are observations or measurements of a variable and a sample is a portion of the population. This is quite a limited aspiration – concentrating on making *predictions* about a defined *population* based only on *measured* data.

Moving now to the practice of statistical modelling, the main sources of criticism of statistical modelling, particularly forecasting, that have been made from a CR perspective are Lawson (1997), Pawson (Pawson and Tilley 1997) Fleetwood (2001), Porpora (1998a), Ron (1999), Olsen (1999), Sayer (1992) and Mingers (2000a). The criticisms are grouped into several themes: the assumed nature of causation within empiricism; assumptions about stability and closure; the ad hoc and a-theoretical nature of statistical modelling; the limitations of the null-hypothesis significance test approach; and the sheer lack of forecasting accuracy.

Empiricism and causation

The implicit philosophy of statistical modelling is inherently empiricist. That is, it largely restricts itself to analysing empirically available quantitative data rather than going beneath the surface to explain the mechanisms that give rise to empirically observable events (Lawson 1997; Mingers 2000a). It embodies a view of causation that is successionist rather than generative, that is based on the Humean conception of constant conjunctions of events. It is assumed that events occur in regular sequences and that these can be effectively quantified yielding a set of related variables, often over time. From these data mathematical equations can be generated that represent semi-universal laws. No other form of causation can be inferred from statistically significant results – they only imply association (Abbott 1998). In contrast, CR emphasizes the importance of generative causation – i.e. moving from observed data associations to the interacting causal mechanisms that underlie them rather than simply a wider population of unobserved data.

This whole approach only makes sense with a particular view of causality. Although this is seldom explicitly discussed, Hendry (Hendry *et al.* 1990, p. 184), for example, accepts that 'I am a Humean in that I believe we cannot perceive necessary connections in reality. All we can do is set up a theoretical model in which we define the word "causality" precisely, as economists do with $y=f(x)$.'

Hendry provides an interesting account of the nature of econometric models:

> There is an economic mechanism which operates in reality: this comprises the transacting, producing, transporting (etc.) behaviour of economic agents

> at some time and place. The data generation process [DGP] superimposes a measurement system on that mechanism, which usually records only part of the information economists require, and may do so with substantial error.... [T]he DGP characterizes the actual stochastic structure of the observables.... A theoretical (economic) model is a conjectured representation of the DGP. Such a model is a cognitive entity ... and generally is contingent on many (usually implicit) restrictions and assumptions about 'everything else held constant'.
>
> (Hendry 1995, p. 55)

This begins by recognizing real economic mechanisms that generate economic events, although in contrast with CR it assumes that reality consists only of individual, economically-rational agents (see intrinsic closure below), ignoring the influence of wider economic and social structures. But it does not make clear the distinction between underlying mechanisms (the Real) and the events that they may *or may not* generate (the Actual). It recognizes the distinction between the intransitive domain of economic reality and the transitive domain of cognitive models. However, the problem is that the Real is immediately reduced to the Empirical with the imposition of a set of measurable and observable variables and the idea of a data generating process that accounts for their statistical characteristics. Even this is subject to severe limitations, as Hendry recognizes, in terms of unobservable variables, immeasurable variables, and major errors of measurement.

I am not arguing here that the situation is different under critical realism. We will always be in the position of attempting to infer unknown mechanisms from limited observations and experiences. The argument is rather that we need to move away from the scientistic obsession with quantifiable data and Humean causality towards a concern with the underlying mechanisms, complex, multidimensional and often unobservable, that, as it were, occasionally give glimpses of themselves in the Empirical domain. There are quantitative, statistical methods that do concern themselves more with trying to discover causal mechanisms from observable data, particularly the work of Pearl (2000) and more generally the Bayesian approach to statistics, which will be discussed below.

Stability and closure

Major assumptions have to be made about the closure and stability of the system under study. Extrinsic closure is the assumption that those factors not included in the model will not change, or will not have a substantive effect on the model if they do. In practice, there will be many variables excluded from the analysis for a variety of reasons. It could be because data is not available, or because the factors are not measurable; or because a factor in not operative at that point but could become so; or because the investigators are ignorant of it. These may well have significant effects on the phenomena being analysed (Liu 1960). Moreover, social systems are never closed but always open to historical change and accident (what is sometimes called 'path dependence', David 1986). Intrinsic closure is the aggregating assumption that economically active individuals are inherently identical, rational, and linearly additive with no emergent properties or non-linear dynamics. A further point here is that the data itself is often extremely unreliable, as much a reflection of its process of production as some real factor. We can illustrate this with Fildes' definition of an econometric model:

> Econometric models are just one of a number of different ways of characterizing an economic or behavioural system. They contrast with 'systems models' in that they are typically aggregative linear or almost linear models with a well-defined stochastic structure. Model parameters are estimated from the data using well-understood and statistically 'optimal' techniques based on these stochastic assumptions. The variables modelled (or their proxies) are typically measurable.
>
> (Fildes 1985: 550)

An example of a 'stochastic assumption' is that the forecast errors are independent, identically normally distributed, random variables. This implies homoscedasticity, i.e. that the error variance is constant across the range of values of the dependent variable. Other assumptions typically made in regression are that all relationships are linear and that the explanatory variables are all normally distributed and independent of each other. We can see immediately the assumptions outlined above concerning stability and closure. It is a given of the whole approach that it makes atomistic and linear aggregating assumptions, and accepts that anything of relevance will be (reliably) measurable.

Ad hoc nature of modelling

For Fildes, having arrived at a set of data, often based on what is available as much as anything else, the next stage is the building of the model itself. Many decisions have to be made in developing a model: its functional form (e.g. linear or non-linear, absolute, differenced, transformed); single equation, structural equations, or vector autoregressive (VAR); what lag structure to use if any; which outliers to delete; what to do with missing data; and which variables to include or exclude.

Here, the major critique is that the model is always under-determined by the data. Many different sets of variables, subject to different sets of relationships, will be compatible with the data. It is generally not possible to choose between them *on purely statistical grounds* and so all kinds of other, judgmental, factors come into play – ad hoc criteria, personal belief or intuition, 'experience', potential usefulness, robustness, etc. These undermine the supposedly 'scientific', observer-independent approach. Often the form of the model is chosen more for its mathematical properties, i.e. its tractability, than its realism and therefore choices are made in an ad hoc way, often based neither on sound statistical theory nor on domain-specific theory.

A further area of significant judgmental adjustment is after forecasts have been made. It is commonplace, particularly among economic forecasters, to adjust forecasts to reflect aspects of the situation not contained in the model, for example unusual data, particular out of the ordinary events, or simply the fact that recent forecasts have under- or overestimated the trend (Fildes and Stekler 2000). Such ad hoc changes do significantly improve at least short-term accuracy, but equally simply illustrate the points made above about model closure, and do little to help longer-term forecasts.

Statistical modelling as a-theoretical

Much statistical and econometric modelling is a-theoretic in that it does not draw on previous economic theory; it simply restricts itself to modelling the data as such. The approach Fildes recommends, 'general to specific' (Hendry 1995), does attempt to draw on previous work, although it is usually previous empirical studies rather than theory directly. But he recognizes that there is a strong tendency to concentrate attention on the data that is available, perhaps using a variety of techniques, rather than trying the harder task of generating new data on the basis of available theories. However, several econometricians are much more critical, arguing that econometrics is, in general, profoundly a-theoretical (Cooley and LeRoy 1985; Hendry *et al.* 1990; Hoover 1988; Koopmans 1947).

Significance testing

Null-hypothesis significance testing (NHST) is a fundamental part of regression methodology – it is used in testing the significance of the regression as a whole (an F-test of the null hypothesis that all the regression coefficients are in fact zero); in evaluating individual regressor variables (a t-test of the null hypothesis that particular coefficients are zero); and in testing the significance of the correlation coefficient. But NHST is much more widely used than that. Indeed, in some disciplines such as psychology and behavioural science it is the *sine qua non* of research methods – statistically significant results are almost obligatory in order for the research to get published in certain journals (Nickerson 2000). However, NHST has been subjected to an enormous amount of criticism in many disciplines and, although it has its defenders (e.g. Chow 1996, 1998; Cortina and Dunlap 1997; Hagen 1997), I will argue that it is now an outmoded approach.

The criticisms of NHST are many and severe:

1 **The logical fallacy.** The logic of NHST rests on *modus tollens* (denying the consequent): '*p* implies *q*; not *q*; therefore not *p*'. 'If H_0 is true the data will follow a particular pattern; the data does not follow that pattern; therefore H_0 is not true'. However, this logic breaks down if the statements are only probabilistically correct: 'if H_0 is true it is likely the data will follow a particular pattern; the data does not follow that pattern; therefore H_0 is unlikely to be true' does not follow, as the example in Cohen (1994) nicely demonstrates.

2 **The inverse probability fallacy.** Many people presume that the smaller the p-value (and thus the more unlikely the actual result) the more unlikely it is that H_0 is true. Or, more generally, that the *p*-value in fact measures the probability of H_0 being true. So, for example if a *p*-value turns out to be 0.04, there is only a 4 per cent chance of H_0 being true. This looks at p the wrong way round. The p-value is the probability of the observed data occurring given that H_0 is actually true – $Pr(D|H_0)$. This is not the same as the probability of H_0 being true give the observed data – $Pr(H_0|D)$. This latter probability depends also upon the prior distribution of probabilities of H_0 and H_1 and the calculated statistic.

3 **The statistical significance fallacy.** One of the major criticisms of work published on the basis of significant results is that it confuses statistical significance with theoretical or practical significance. A test may demonstrate statistically that there is evidence that the observed effect is unlikely to have occurred by chance. But that says nothing about the magnitude, importance, or theoretical significance of the effect. McClosky and Ziliak (1996) conducted a survey of all papers published in one of the leading economic journals (*American Economic Review*) during the 1980s that employed regression analysis. They found that all the papers reported the *statistical* significance of their results but that 70 per cent did not discuss the *economic* significance either for policy or for theory. 50 per cent did not even discuss the magnitude of the effects they were reporting.

4 **The sample size fallacy.** It is commonly believed that the larger the sample size the more reliable the result, especially when rejecting H_0. In fact, the sample size does not affect the significance level (α) – the probability of a Type I error – at all. What it does affect is β (the probability of a Type II error) and therefore the power of the test. In particular, the larger the sample the smaller any difference in magnitude that will be detectable for a particular value of α. Given a large enough sample, even a very small difference, and there will always be some observed difference, will become significant. Thus a large sample size, rather than being beneficial, leads to the detection of often spurious and certainly theoretically insignificant differences or effect sizes.

5 **The 'nil' hypothesis problem.** The null hypothesis actual got its name as the hypothesis to be 'nullified' but it is generally taken to be null in the sense of zero, no change or no difference, what Cohen (1994) calls the 'nil hypothesis'. As such, it is the case that H_0 can never in reality be true. There will always be some dimension or level or precision at which the two populations will differ and, given a large enough sample, such differences will be detected. Furthermore, what does it mean if H_0 is not rejected? This does not prove that H_0 is true, and indeed it almost certainly could not be. It can only say that we do not have sufficient evidence to reject a hypothesis we know to be wrong anyway! It thus provides no useful information whatsoever. What if we do reject H_0? The only information that this provides is that there is some evidence for change or for the direction of change but nothing concerning the magnitude or practical significance of any effect.

6 **The problem of the arbitrariness of α.** It has become conventional to set α at particular pre-defined levels such as 5 per cent or 1 per cent. This is obviously quite arbitrary and stems originally from the necessity to hand-calculate statistical tables. Worse than arbitrariness is the all-or-nothing result: 4.9 per cent is significant – hooray, publish – 5.1 per cent is not. In practice, in my experience, results tend to be either clearly significant, way above critical values; clearly insignificant, with obviously no effect present; or in the grey areas in between. In the first two cases the test is hardly necessary, and in the latter case it is of no use.

7 **The biased research problem.** One 'significant' effect of the emphasis on significance testing in research is that it leads to biases in the research literature. The pressure to only publish (statistically) significant results clearly leads to a bias against accepting the null hypothesis. This can have two effects: *theoretically* significant results rejecting alternative hypotheses or theories (i.e. *statistically* insignificant) do not get published; and the results that do get published will include an inflation of Type I errors. This latter problem will result either from researchers increasing their sample size until they do get a significant result – see the sample size fallacy above – or simply because the occurrences of a wrongly rejected H_0 will be a much greater proportion of the actually published studies.

Given these damning criticisms, can anything be rescued from statistical testing? Several responses have been proposed, some of which critical realism would approve of. Unfortunately, there is little evidence in, say, the information systems literature of this being put into practice. The first is using the effect size and interval estimation instead of testing. There are various possibilities, but the approach involves abandoning hypothesis tests and instead reporting the actual size of the effect, in terms of the calculated coefficients, together with confidence intervals around them (Casella and Berger 1990; Gardner and Altman 1986; Meehl 1997). This approach avoids several of the problems with hypothesis testing discussed above and it focuses attention on the actual magnitude of the effect and thereby on its practical significance.

However, confidence intervals (CIs) are still based on arbitrary values of α – should it be 95 per cent, 99 per cent or what? And they still rest on the assumptions underlying significance testing, which are often not valid. Moreover, and this is possibly an explanation for the lack of take-up, the intervals are generally embarrassingly wide for the sample sizes and sample variances that commonly occur. CIs are also easy to misinterpret – properly, a 95 per cent CI based on a sample mean implies that 95 per cent of such intervals will contain the true mean (which is actually fixed but unknown). It is often interpreted to mean that there is a 95 per cent chance that the true mean lies within the calculated interval. Reichardt and Gollob (1997) suggest several other reasons for their unpopularity.

Forecasting accuracy

The final criticism, a consequence of the above points, is simply the abysmal performance of such forecasting methods in practice. At a general level there is some agreement that econometric models do very poorly in ex-ante forecasting (Hutchison 1994; Kay 1995; Leamer 1983; Rosenberg 1992; Smith 1995): that is, forecasts made before the event, using where necessary forecasts of the independent variables, as opposed to ex-post forecasts using the actual values of the independent variables. Even in the latter case, performance on holdout samples (i.e. samples of data not used in estimating the model) is much worse than within the sample on which the model was estimated.

Sherden (1998), in a provocative book about the difficulty of predicting the future, has documented how poorly forecasters in a whole variety of fields perform: 'Of these sixteen types of forecasts, only two – one-day-ahead weather forecasts and the aging of the population – can be counted on. The rest are about as reliable as the fifty-fifty odds in flipping a coin' (p. iii). Even Fildes (2000), in a partly critical review of the book, accepts that much of the criticism made of economic forecasters is justified.

I can perhaps summarize the state of play with Allen and Fildes' conclusion: 'A well specified econometric model should forecast at least as well as the naïve no-change method.' (2000: 348). Given the problems, outlined above, of the availability and reliability of data, the difficulty of developing a 'well-specified' model, and the significant costs in time and expertise involved, then this is surely damning with faint praise.

9.2.3 Bayesian modelling

As a slight digression, we want to consider briefly an approach to statistical modelling which is an alternative to the traditional one discussed above. Although it often ends up in the same place, it approaches from an interestingly different starting point and seems to have a greater affinity with the critical realist approach. In contrast to the traditional approach, Bayesian modelling is characterized by the following (French and Smith 1997; West and Harrison 1989; Zellner 1997):

- An interest in the subjective knowledge, beliefs and judgments of those concerned with a particular situation or decision. These beliefs, which will generally be uncertain, can be represented by *subjective* probabilities.
- An explicit approach to modelling such beliefs that allows both changes in belief in the light of new information, and the combining together of information from different sources.
- A focus on *causal structures* and particularly the updating of beliefs about underlying causes in response to the occurrence (or non-occurrence) of particular events.

Technically, this approach stems from a well-accepted mathematical result called Bayes' theorem which is used to adjust probabilities in the light of new data. Suppose that we have a die but are not sure if it is biased towards a six or not. Our *prior* probabilities are, say, 50/50 that it is unbiased. We throw it four times and get three sixes. This outcome or event can be used to adjust our view about the die. We calculate the *likelihood* of this outcome given the two possible situations or causes. It is clearly more likely to have occurred if the die is biased, and we can use Bayes' theorem to update the prior probabilities into *posterior* ones – say 60/40 in favour of a biased die. This process could be continued with more data generating more certainty about the situation.

There is no debate about this procedure from a technical point of view – Bayes' theorem follows from the laws of probability and is accepted by traditionalists. Debates come in terms of the source of the probabilities. Traditional statisticians generally take a 'frequentist' view that valid probabilities can only come from the objective observation of relative frequencies and deny the legitimacy of subjective probabilities. Bayesians, in contrast, are very happy to use subjective judgment and expert knowledge along with empirical data where it is available. Bayesian modelling has a very wide range of potential application areas – decision analysis and decision-making in general (French and Smith 1997), forecasting (Pole *et al.* 1994), traditional statistics (Berry 1996), and belief networks (McNaught 2001; Oliver and Smith 1990). The latter are particularly interesting from a critical realist perspective as they can be seen as a way of operationalizing CR's retroductive methodology.

Pearl has also developed a method that aims to uncover underlying causal relationships from statistical data rather than remaining at the associational or probabilistic level – see Section 5.5.1.

9.3 Qualitative methods in systems research

Qualitative methods are those that generate information of a non-measurable nature. This includes traditional methods such as interviews, ethnography, hermeneutics, participant observation and also systems methods oriented towards action such as soft systems methodology (SSM), cognitive mapping, action research, the viable systems method (VSM) and so on. Generally speaking these are underwritten by interpretivist philosophical assumptions that stress the meaningful nature of social processes and the need to understand, and see things from the point of view of, participants in the research situation.

There is no problem with these research methods in principle; they are obviously valuable and necessary for understanding the processes of the social world. The problem comes about philosophically when the proponents of them adopt a strongly anti-realist position that either denies the existence of a social reality above and beyond individual meanings (what Bhaskar calls the epistemic fallacy); or denies that we can have access to such a world; or claims that different viewpoints, whether of the layman or the theorist, are all equally valid (judgemental relativity). The flavour of these arguments can be seen from a revealing debate within the management literature. Meckler and Baillie (2003a, p. 274) put forward 'a middle-way between the postmodern rejection of notions of truth and objectivity and the strong positivist position', which was not too far distant from critical realism. This met with a vituperative response from Gioia (2003) and Lounsbury (2003), generating a reply from Meckler and Baillie (2003b) and a supportive commentary from Hunt (2005).

As with quantitative methods, there is a vast range of qualitative methods and so I will confine myself to discussing two soft systems approaches in particular: cognitive mapping and SSM, which are two of the most widely used (Rosenhead and Mingers 2001).

Cognitive mapping is a diagrammatic technique for depicting the way an individual thinks about a particular issue or problem (Bryson *et al.* 2004). In appearance it is similar to an influence diagram, but it clearly aims to map *a person's beliefs about* an issue, rather than 'objective' aspects of the situation. It is primarily used within strategic decision-making as part of a wider process that was known as SODA (Eden 1989) and is now called Journeymaking (Eden and Ackerman 1998). Cognitive mapping is based on Kelly's psychological theory of constructs and is clearly subjectivist in limiting itself to exploring people's beliefs about the world. It can, therefore, clearly be seen as interpretive or hermeneutic in character and as such in opposition to the intransitive world of real structures and objects.

However, I would argue that this is not wholly the case. It is part of a wider process within which the actor-independent world is considered. Different people's individual maps can be compared and, through discussion, group maps can be developed. This process becomes less and less subjective, and can result in substantive, real-world research. 'It is usual for a SODA workshop to identify opportunities for further analysis, such as financial model building, simulation modelling, market research, and statistical analysis' (Eden 1989, p. 39). In one documented case (Bennett *et al.* 1997), cognitive mapping was combined with a system dynamics model that was explicitly aiming to be demonstrably valid in depicting actual occurrences. Overall, it is better seen as a qualitative component within a pluralist research framework such as multimethodology (Mingers and Gill 1997).

Within soft systems, Checkland (1999b; Checkland and Holwell 1998; Checkland and Poulter 2006; Checkland and Scholes 1990; Mingers 2000b) and his development of soft systems methodology (SSM) most clearly articulate a philosophy based on phenomenology:

> [We] need to remind ourselves that we have no access to what the world *is*, to ontology, only to descriptions of the world ... that is to say, epistemology.... Thus systems thinking is only an epistemology, a particular way of describing the world. It does not tell us what the world *is*. Hence, strictly speaking, we should never say of something in the world 'It is a system', only: 'It may be described as a system'.
>
> (Checkland 1983, p. 671)

This stance has involved a strong ontological move, denying that 'systems' have reality in the world and allowing the term only epistemological validity as a way of helping our thinking: 'Systemicity is shifted from the world to the process of enquiry into the world: "the system" is no longer some part of the world which is to be engineered or optimised, "the system" is the process of enquiry itself.' (Checkland and Scholes 1990, p. 277.) As these quotations show, Checkland denies the ontological reality of 'systems', instead reserving this concept for *thinking about* the world. He also distinguishes strongly between natural and social science, or rather positivist and phenomenological approaches within social science, and allies SSM clearly with the phenomenological tradition. I shall have to restrict myself to making a few observations on SSM from a CR perspective. The main problem is that Checkland takes positivism as the only alternative to interpretivism as a philosophy of (social) science. This inevitably means that he has to adopt a full-blown phenomenological position that then generates all kinds of contradictions and problems in dealing with a 'real-world' external to the observer that is, after all, what SSM aims to improve (Jackson 1982; Mingers 1984). The major advantage of a critical realist approach is that it maintains reality while still recognizing the inherent meaningfulness of social interaction.

It might be said that SSM only concerns ideas or concepts (root definitions or conceptual models) and that these are somehow less real than objects; or, that it is strongly relativist in accepting all viewpoints as being equally valid. Against this, critical realism demonstrates that ideas, concepts, meanings and categories are as real as physical objects (Bhaskar 1997). They are emergent from, but irreducible to, the physical world, and have causal effect both on the physical world (e.g. in the generation of technology) and the social and ideational world. They are also inevitably social products and participate in transformations of the social world, just the sort of transformations that SSM aims to bring about. With regard to relativism, CR makes a distinction between epistemic relativism and judgmental relativism – people may well hold different beliefs about processes in the world, but this does not mean that we are unable to rationally judge between them and prefer one to another given some particular purpose. Equally, ideas once expressed are no longer wholly subjective – they become intransitive and available for investigation, debate and judgement by others. This is an example of a more general idea – referential detachment (Bhaskar 1994, p. 52) – that any communication must refer to something, that which it is about (even if it is self-referential), and this immediately establishes an intransitive dimension. Bhaskar goes further in arguing against the positivist distinction between facts and values (which would fit in well with both soft and critical systems), and eventually to a moral realism – i.e. the idea that there could be moral truths (Bhaskar 1994, p. 108); see Chapter 10 for further discussion.

A final point is the weakness of SSM with regard to the origin of the world-views that it explores, and an understanding of the difficulties of individual and organizational change. These both stem from the individualistic social theory that SSM embodies. With a critical realist interpretation both of these are avoided. On the one hand we can generate explanations of why particular actors may hold the beliefs they do in terms of their social and organizational position; their history of experiences particularly as these relate to underlying social characteristics such as gender, race, and age; and, of course, their individual personalities (Whittington 1992). We are also in a position to understand the psychological and social structures that may impede or facilitate learning and change.

9.4 Multimethodology

For many years, during the 1980s and 90s, there was a huge gulf between positivism and interpretivism underpinned by a strong belief in paradigm incommensurability as emphasized by Burrell and Morgan (1979). Research had to be either one or the other, and the idea of mixing methods from different paradigms was seen as anathema. Gradually, however, views changed, mainly because of recognition that the world was too complex to fit neatly into only one of the boxes and that real-world research necessitated a combination of qualitative and quantitative data. Mingers and Gill (1997) published a book on 'multimethodology' in 1997; Tashakkori and Teddlie (1998) published their seminal guide to mixed methods research in 1998 and their handbook of mixed methods in 2003 (Tashakkori and Teddlie 2003); the *Journal of Mixed Methods Research* and the *International Journal of Multiple Research Approaches* both started in 2007; and the Mixed Methods International Research Association has been formed. So, although there are still dyed-in-the-wool positivists or constructivists it is now arguable that multimethodology is becoming the norm (Onwuegbuzie 2012).

The fundamental tenet of multimethodology is the importance of combining together methods or sometimes methodologies in dealing with real-world situations, whether the purpose is pure research or a practical intervention. (In fact these are not a dichotomy but ends of a dimension. All research involves some form of intervention or effect, and all interventions involve research.)

The first argument in favour of multimethodology is that the real world is complex and multi-dimensional, while particular research or intervention methodologies focus only on specific aspects. Using a particular methodology is like viewing the world through a specific instrument such as a telescope, an X-ray machine, or an electron microscope. Each reveals certain aspects but is blind to others. Although they may be pointing at the same place, each instrument produces a different, and sometimes seemingly incompatible, representation. Thus, in adopting only one method one is often gaining only a limited view of a particular research situation: for example, attending only to that which may be measured or quantified or only to individuals' subjective meanings and thus ignoring the wider social and political context. This argument is a strong one in support of multi-method research, suggesting that it is always wise to utilize a variety of approaches.

A framework has been developed (Mingers and Brocklesby 1997) from Habermas's (1979, 1984, 1987, 1993b) theory of communicative action which highlights three 'worlds'. These are the *material world* of actual and possible states of affairs; the *social world* of normatively regulated social relations and interactions; and the *personal world* of experiences and beliefs (see Midgley 1992 for a similar approach) – see Figure 9.1.

The second argument for combining methods is that research and intervention is not a discrete event but a process that has phases (or, rather, different types of activities) that will predominate at different times. Particular methodologies and techniques are more useful for some functions than others, and so a combination of approaches may be valuable to provide a more comprehensive outcome. An analysis of the general process of research and intervention, based on work of Bhaskar (1979: 144), Maturana (1990: 18), and Tashakkori and Teddlie (1998), led to the identification of the following four generic phases (Mingers and Brocklesby 1997):

- **Appreciation** of the research situation as experienced by the researchers involved and expressed by any actors in the situation, and prior literature and theories.
- **Analysis** of the information from the first stage so as to understand the history that has generated it, and the particular structure of relations and

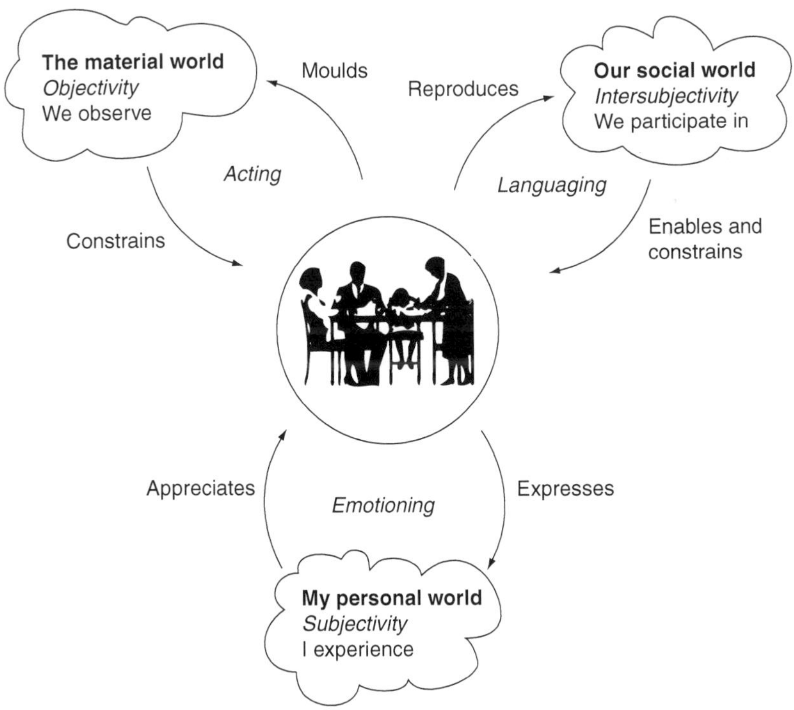

Figure 9.1 Habermas's three worlds.

constraints that maintain it. This will involve analysis methods appropriate to the methodology of the study and the data produced in the first stage. Explanation will be in terms of possible hypothetical mechanisms or structures that, if they existed, would produce the phenomenon that has been observed, measured, or experienced.

- **Assessment** of the postulated explanation(s) in terms of other predicted effects, alternative possible explanations, and, within action research, consideration of ways in which the situation could be other than it is.
- **Action** to bring about changes if necessary or desired, or to report on and disseminate the research results.

Put crudely, these phases cover: What is happening? Why it is happening? How could the situation or explanation be different? And, how could the difference be made real? We should emphasize immediately that these activities are not seen as discrete stages that are enacted one by one. Rather, they are aspects of the intervention that need to be considered throughout, although their relative importance will differ as the project progresses. Equally, different studies will place their emphasis at some stages rather than others.

These two factors – the multidimensionality of real-world situations and the different types of activity that need to be undertaken – can be combined to produce a grid (see Table 9.1) that can be used to map the characteristics of different research methodologies to help in linking them together.

The logic of this framework is that a fully comprehensive research intervention needs to be concerned with the three different worlds – material, personal, and social – and the four different phases. Thus each box generates questions about particular aspects of the research situation that need to be addressed. This framework can be used in two ways. First, it can be used to map particular methods or methodologies in terms of where their strengths lie – some may be very specific to a particular cell or two, others (e.g. SSM) may be useful across many cells. A map is produced for each method(ology)

Table 9.1 Framework for mapping methodologies

	Appreciation of	*Analysis of*	*Assessment of*	*Action to*
Social	social practices, power relations	distortions, conflicts, interests	ways of altering existing structures	generate empowerment and enlightenment
Personal	individual beliefs, meanings, emotions	differing perceptions and personal rationality	alternative conceptualizations and constructions	generate accommodation and consensus
Material	physical circumstances	underlying causal structure	alternative physical and structural arrangements	select and implement best alternatives

showing its strength in a cell by a particular colour. This mapping has been done for a range of management systems methodologies in Mingers (2003). These maps can also be combined to show all the methods are suitable for a particular cell: see Table 9.2.

Second, the framework can be used to analyse the tasks in a research/intervention project to see what the requirements for methods will be. Although in principle all the cells might be relevant, in practice some may be more important than others in a particular intervention. For example, a project concerning logistics or production processes might be dominated by physical issues, and there may be high-quality data available which would particularly involve the 'material world' cells. This is not to say that social, and even personal, issues are irrelevant but to accept that they may be second-order considerations. Equally, a problem may revolve around differences in values or viewpoints, and quantitative data may be scarce or unreliable, thus highlighting the 'social world' boxes.

A third argument in favour of mixed methods is that it encourages triangulation of research and generates more interesting and stimulating results. The idea of triangulation (Modell 2009), which comes from surveyors using three different lines of sight to fix a position, is to employ a variety of research methods, sources or researchers in order to enhance the validity of the research findings, particularly by reducing the biases of any one approach. In line with the arguments above, different research methods produce different views of the same phenomena. The results so produced may be similar, in which case they provide greater confidence in the conclusions, or they may differ, in which case the contrasts generate interesting questions for further analysis and may well lead to a greater understanding of the complexities of the phenomena.

In terms of the actual processes or mechanisms for combining methods, there is now a range of materials available in the literature. The Tashakkori and Teddlie books mentioned above are invaluable. There are also now a number of research methods textbooks and papers that have been written specifically from a mixed methods perspective and should enable everyone to combine together quantitative and qualitative research methods in successful research and intervention. Examples are: Bryman (2006); Creswell (2009); Greene (2007); Plano *et al.* (2007); Plowright (2011); Venkatesh *et al.* (2013).

9.5 Critical realism and research

As was mentioned in section 9.2.2, critical realism has traditionally been quite critical of quantitative methods because of the untenable empiricist assumptions that they presume, and that has led to the belief that CR primarily adopts interpretive research methods. However, I wish to argue that that is not in fact the case and that CR actually licenses a multimethodological approach. I will then go on to argue that CR also brings with it particular principles – e.g. abduction, explanatory mechanisms – that enhance the basic multimethodology position.[1]

Table 9.2 Mapping of research methods

	Appreciation of current situation	*Analysis of underlying structure*	*Assessment of possible alternatives*	*Action to change*
Social world	Actor-network theory Critical theory Critical systems heuristics Dialectical hermeneutics Ethnography Grounded theory Interviews Participant-observation Social statistics Soft systems methodology (SSM)	Actor-network theory Critical theory Critical systems heuristics Dialectical hermeneutics Grounded theory Interviews Participant observation Social statistics	Actor-network theory Critical systems heuristics Critical theory Dialectical hermeneutics Language/action approach	Critical systems heuristics Dialectical hermeneutics Language/action approach Social statistics Viable systems model
Personal world	Cognitive mapping Hermeneutic/interpretive Interviews Participant observation Phenomenology Questionnaire Soft systems methodology	Cognitive mapping Hermeneutic/interpretive Interviews Participant observation Soft systems methodology	Case study Role play Soft systems methodology	Case study Personal development workshops Soft systems methodology
Material world	Measurement Quantitative data production Questionnaire/survey Statistical analysis	Experiments – field Experiments – laboratory Simulation Statistical analysis Systems dynamics modelling Viable systems model	Cost-benefit analysis Experiments – field Experiments – laboratory Simulation Systems dynamics modelling Viable systems model	Viable systems model Simulation Statistical analysis/presentation

9.5.1 Critical realism and multimethodology

We should also mention that, prior to critical realism coming to the fore, critical theory, especially based on Habermas's work, also accepted a variety of methodological approaches although it did not perhaps strongly advocate their combination. Habermas's first major theory was known as the theory of knowledge-constitutive interests (Habermas 1978) and recognized three forms of knowledge based on three underlying human interests – empirical science (technical interest in control of nature); hermeneutic science (practical interest in human communication); and critical science (emancipatory interest in freedom from dogmatism). These formed the basis of critical systems thinking (Flood and Jackson 1991; Jackson 2003; Midgley 2000; Mingers 1997c), which applied these ideas to using systems methodologies in practical problem-solving interventions (known as hard, soft and critical). See also Morrow and Brown (1994) for a book on critical research methods and Myers and Klein (2011) for a set of research principles.

In essence, CR is highly pluralist in terms of both ontology and epistemology. It recognizes the existence of a whole range of entities – material objects and forces; social structures and practices; conceptual systems such as languages, beliefs and reasons; and feelings and experiences. All of these are 'real' because they have causal efficacy, even though they may not be observable or perceptual (Sayer 1992, 2000). These entities do differ, however, in our means of access to them: i.e. epistemologically. Physical objects can be observed and measured; social norms and practices require qualitative investigation perhaps from a participant perspective, while personal values and feelings require hermeneutic or phenomenological analysis. Critical realism is therefore happy to accept the validity of a wide range of research methods without recognizing the primacy of any particular type or approach (Angus and Clark 2012; Danermark *et al.* 2002; Houston 2010; Miller and Tsang 2011; Pratschke 2003; Zachariadis *et al.* 2013).[2]

While CR recognizes that these methods are useful in principle, a researcher has to be very much aware of the assumptions and limitations of these approaches. In other words, it is not satisfactory to use quantitative methods, for example, in a naively empiricist way, as this would be subject to all the criticisms outlined in section 9.2. It has to be recognized that the goal of the research is explanation rather than mere description or summarization; that attempts must be made to go beneath the empirical level to search for explanatory mechanisms that might generate the data; that the social world is inevitably open and unpredictable; and that research is never value-free but always committed to a particular purpose or interest (see the extended discussion in section 10.5).

Equally, with qualitative or interpretive methods it has to be accepted that there is an independent, objective world even if access to it is always socially mediated; that we should not commit the epistemic fallacy of assuming that limitations to knowledge are limitations to being itself; that we do not have to accept that different viewpoints are equally valid (judgemental relativity); and that social science is inevitably committed to truth and revelation rather than simple description.

9.5.2 Critical realism's research principles

Beyond the support for multiple methods, to properly carry out critical realist research one should also follow principles that follow from CR's basic tenets, and particularly its own methodology – DREIC – Describe, Retroduct, Eliminate, Identify and Correct (see Section 2.3.2). For applied research, this has been expanded to RRREIC (Bhaskar 2010, 2013):

- *Resolution* of complex phenomena into components
- *Redescription* in an explanatory meaningful way
- *Retroduction* of potential hypothetical explanatory mechanisms
- *or Retrodiction* of potential antecedent causal events
- *Elimination* of alternative competing explanations
- *Identification* of causally efficacious mechanisms
- *Correction* of earlier findings/theories

The difference between retroduction and retrodiction was mentioned in Section 5.2.2 (Bhaskar 1994, p. 30). With retrodiction we are hypothesizing possible antecedent causal *events*, while with retroduction we are looking for the underlying generative mechanisms. Of course, in actual applied research we will need to consider both. CR also (Bhaskar 2010) recognizes that in the real world open systems are complex and interactive, operating at different levels and also within different disciplines – typically psychological, economic, social, political and perhaps also physical. Bhaskar refers to this as a 'laminated totality'. One of the implications of this is that in applied as opposed to pure, theoretical research there is a fundamental need for interdisciplinarity which complements the methodological pluralism discussed above.

Critical realist research should also avail itself of the categories and distinctions within CR:

- Research must pay attention both to the object/problem it studies (intransitive) and the place/context of the study (transitive).
- It should involve distinctions between
    - structure/events;
    - Real/Actual/Empirical;
    - transfactuality of laws.
- Social phenomena should involve
    - the use of the TMSA (further developed by Archer);
    - social life as concept-dependent but still intransitive;
    - the use of the MELD categories – being as structure, being as process, being as totality, being as human agency.

Wynn and Williams (2012) identify five fundamental principles: identification of particular events to be explained; identification of relevant context and structure (both social and material); retroduction (or abduction) to potential generative mechanisms; empirical corroboration to establish causal powers and decide between alternatives; and the use of triangulation and multimethods.

To begin with there must be something to be explained, something that is to be the object of the knowledge generated by the research. This is generally a series of events that are experienced by participants and observers. In order for them to be interesting in research terms, they must be in some way unexpected, either in terms of everyday experience, or in terms of the existing theories in the area. Moreover, events may be positive things that happen, in which case the question is: what is causing them to happen? Or they may be absences – things that are expected to happen but do not, in which case the question is: what is preventing them from happening? Both are equally important. At this stage one may use any of the research methods discussed above to pick the events of significance out of the continual flux of experience. They may be based on observation, measurement, inteviews, ethnography, participation etc., and may come from both researchers and actors. This is essentially the 'appreciation' phase discussed in Section 9.4.

The next stage is to begin analysing the structures and contexts that are relevant to explaining the events. This is very much a *systems* analysis, picking out systems, components, relationships and processes that are active in the context. They may be physical, social, psychological, informational or economic. It is important to understand the causal powers (or emergent properties) of these systems, and how they interact with each other. This is the 'analysis' stage from Section 9.4.

The third stage is really part of the previous one. Hypotheses are generated as to which generative mechanisms might be behaving or interacting in such a way as to produce the events in question. Equally, if the events are in fact absences, it may be a question of which mechanisms are countervailing others to prevent the expected events occurring. This is very much a time for insight and imagination. Whereas the earlier stages can be performed in a fairly rigorous and prescribed manner, this is more a question of discovery rather than justification. At the end of this phase, there will probably be several competing hypothetical explanations, or perhaps just one that is in need of justification and corroboration.

The fourth stage is one of corroboration and justification, what was called 'assessment' of alternatives in Section 9.4. We need to gather evidence and test the various competing explanations. This may well involve significant research to evaluate the strength of the claims of the explanations, perhaps involving predictions or statistical analysis of experimental data or follow-up interviews. The aim is to select the mechanism that offers the best and strongest explanatory power. Where the aim of the research is some form of intervention or action research, this stage can also include consideration of possible changes to the situation that would bring about preferred outcomes. In this situation, the 'action' phase then needs to be considered.

9.5.3 An illustrative example

In this section I will provide a brief example to illustrate the principles mentioned above. The example actually comes from Malcolm Gladwell's (2001) excellent book, *The Tipping Point*, and concerns the number of cases of sexually transmitted diseases (STD) in Baltimore in the mid-1990s. For many years, the number of reported cases remained very constant around 200 per year, but then between 1993 and 1997 it rose to over 700, an increase of more than 300 per cent. What could have caused this dramatic increase?

Here clearly are the events to be explained – the sudden dramatic rise in STD – although of course with more detailed information than I am giving here.

The next stage is to analyse the components and contexts of the situation. In this case, there will be factors that generate new cases of STD, and at the same time there will be factors that reduce it. A useful way of representing this is with an influence diagram (also known as a multiple cause diagram): see Figure 9.2.

We can see at the centre two causal loops. The top one is a positive or reinforcing loop that drives up the number of infected people. The more infected people there are, the more new infections that emerge, dependent on the rate of exposure and the rate of infection. Underneath is a negative or balancing

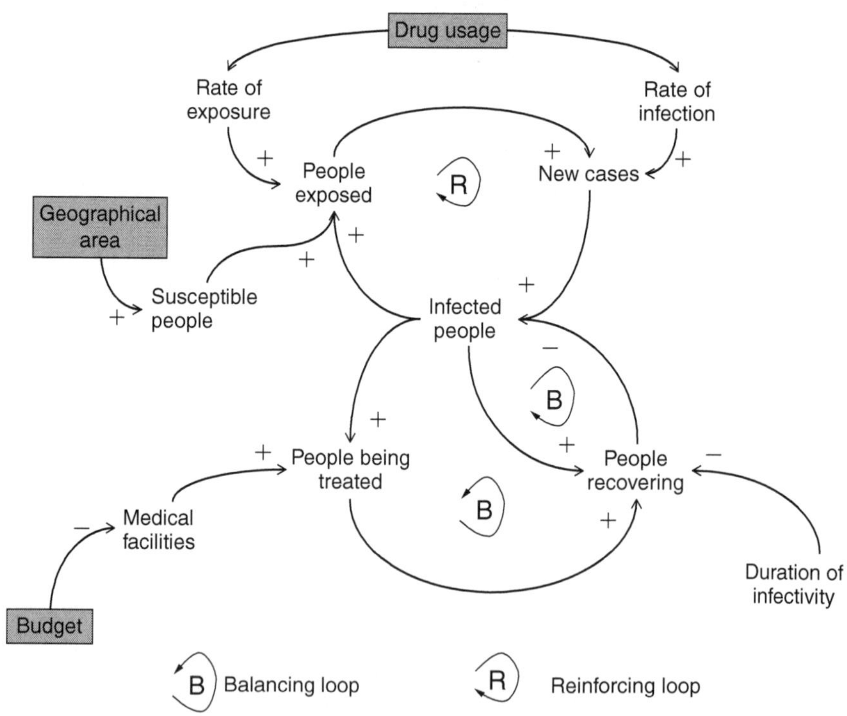

Figure 9.2 Influence diagram of Baltimore STD generative mechanisms.

loop where the new infections are reduced, partly by people recovering, and partly by them being treated medically. While the number of infections was constant, the two loops were balancing each other out. This is a good example of absence or no change. The figures remained the same not because nothing was happening, but because two mechanisms were effectively cancelling each other out.

We now come to the third stage: explanations for the changed events. As it turned out, three different explanations were proposed. The Center for Disease Control (CDC; Center for Disease Control 1996), an official government body, proposed that the explanation was the increased use of crack cocaine. This both increases risky sexual behaviour and brings more people into poorer areas to obtain the drug. In Figure 9.2 this is shown to affect both the rate of exposure (more sexual contact) and the rate of infection (more likely to become infected).

However, Dr Jonathan Zenilman, from John Hopkins Medical School in Baltimore, disagreed. He argued that the primary reason was a cutback in medical services at the STD clinics because of cuts in the city's budget. The number of medical staff went from 20 to ten and patient visits dropped from 36,000 to 21,000. People were cured less quickly, and so infected more people. In Figure 9.2 this reduces the number of people being treated and so reduces the effectiveness of the control loop.

Finally, another eminent epidemiologist, John Potterat from El Paso Department of Health, suggested that it was actually caused by the city's slum clearance programme, which had been operating during this period. The slums in the very poor districts had been the centre of STD infection. As they were demolished, their inhabitants were moved elsewhere throughout the city and so the syphilis that they brought with them was spread much more widely. In Figure 9.2 we can see that this increases the number of susceptible people who can be infected.

So here we have three plausible mechanisms, any one (or combination) of which could potentially have generated the observed events. We can also see the two forms of causality in play – the actual (diachronic) causality from one event to another, and the real (synchronic) causality of generative mechanisms.

The final stage is to carry out further research to either eliminate some of the possibilities, or establish more confirmatory evidence. Such research could involve observing patterns of sexual behaviour, interviewing those involved, or mapping out movements. System dynamics could be used to simulate the system (based on the influence diagram) and see how large the effect strengths would need to be to generate the observed behaviour. Perhaps no one of the mechanisms could do it, only two or three together. Potterat, in a study of another city, found that out of a population of 100,000, a major proportion of the infections were actually caused by only 168 individuals living in a small area and frequenting the same six bars. In the actual example this stage was not carried out as such, but changes to the situation were brought in, especially more education and awareness training and more medical staff, which did eventually bring the epidemic back under control.

9.6 Conclusions

In this chapter we have described an approach to practical research that draws on all three of the themes of the book – systems thinking, mechanisms, and critical realism. In the first part of the chapter we developed critiques of both quantitative and qualitative research approaches, particularly to the extent that they are viewed as the only valid ways of producing knowledge. From this, we argued for the necessity of mixed methods research (multimethodology) wherein a range of different methods, from different philosophical paradigms, were combined together in a research study. This was shown to be an approach that followed from the basic principles of critical realism. Finally, we elaborated on the research principles inherent in CR beyond multimethodology, particularly the importance of explanation rather than just description; the importance of generative causal mechanisms; and the use of the principle of induction (retroduction) beyond induction or deduction.

This chapter has elucidated what it is *possible* to do in research terms; then next chapter considers what it is *desirable* to do, from an ethical or moral standpoint, and discusses whether facts can be separated from values, and whether social science can be value-free or must ultimately be committed social change.

Notes

1 For a valuable paper outlining principles for conduction CR case study research, see Wynn and Williams (2012). See also a special issue of *MIS Quarterly* (37:3, 2013) on critical realism for several empirical examples within the field of information systems.

2 There was a special issue of *Philosophica* (71, 2003) entitled 'Critical realism and methodological pluralism in the social sciences'.

10 Ethics

What *should* we do?

10.1 Introduction

Much of this book has been concerned with fairly abstract and philosophical matters but the last chapter moved us into the world of action, at least insofar as research is concerned. And it has always been my view that knowledge and action were in a sense inseparable. Any action implies some form of knowledge, even if it is implicit or embodied and beneath the consciousness of the person involved. Equally, knowledge about how the world is always has implications for action that we may take. This is akin to Foucault's view that knowledge and power are so inextricably linked that he always refers to power/knowledge (Foucault 1980).

So, I think it is vital, especially at this time of major world crises – environmental, economic, political, ethnic and religious – that we ask not just what our knowledge and technology *allows* us to do, but what in fact we *should* do from an ethical or moral standpoint. Clearly moral philosophy is a huge subject and not one that could be adequately covered in a single chapter, but it is, in fact, at the heart of critical realism, being one of the reasons it is called 'critical' (Bhaskar 2002b). At the same time, Habermas (1992b) (who we have previously identified as a major, systems-based sociologist) has developed a very innovative and influential approach that is known as 'discourse ethics'. This has resonances with, as well as differences from, both soft systems methodology (Checkland 1999a) and critical realism. So the core of this chapter will be a discussion of ethics through a detailed comparison of discourse ethics and critical realist ethics. In order to give this some context, we will ground the discussion in the ethics of business activity, clearly of major significance in the aftermath of the world financial crisis.

10.2 Philosophical ethics

Ethics has been an abiding question within philosophy going back to the Greeks and beyond. In more modern times, Kant is seen as the major figure and other theories can best be described in terms of their relationship to Kantian *deontology* (coming from the Greek for duty). I will outline what are seen as the three major positions within ethics – first, utilitarianism/consequentialism; second, deontology; and third, virtue ethics and communitarianism (Baron *et al.* 1997; LaFollette 2007; Singer 1993). As we shall see, discourse ethics encompasses elements of all three.

One of the principal distinctions in ethics is whether an act should be judged as an act in itself or in terms of its effects and consequences. Consequentialists such as David Hume (1967, orig. 1750) and Adam Smith (2002, orig. 1759) held that proper actions are those that do the greatest overall good or the least overall harm. This was developed as utilitarianism by social reformers such as Jeremy Bentham (1948, orig. 1789), who wanted to displace traditional duties and religious rules with actions that would genuinely improve peoples' lives. Thus 'good' actions are those that bring about the most good for the most people.

Kant (1991, orig. 1785) developed his own theory in direct response to the utilitarians. For him, actions should be seen as morally right or wrong, just or unjust, in themselves regardless of their consequences or the extent to which they benefit particular people. He developed a principle, the 'categorical imperative' (Kant 1991, orig. 1785), which should be followed by all people at all times. The underlying argument for this is that most actions are done to achieve a purpose – they are means to an end, and it is the end that is valued. However, people may value different ends or objectives so can there ever be a universal end that everyone would agree to? Kant's answer was that there could be – human beings in themselves. It is rational human beings who make value judgments and so we need to treat other humans as equal to ourselves, as ends and not means. He formulated this principle in several ways: 'Act only on that maxim through which you can at the same time will that it should become a universal law.' (Kant 1991, orig. 1785, p. 97.) 'Act in such a way that you always treat humanity, whether in your own person or in the person of any other, never simply as a means, but always at the same time as an end.' (p. 106.) So there are two primary aspects to the categorical imperative: that moral behaviour always involves treating people equally as ends in themselves, never as means to an end; and that action maxims should be those that can apply universally, to all people at all times. We shall see both of these ideas being embodied in discourse ethics, although in a very different way.

While Kant's theory was quite individualistic – it is the individual, rational subject who has to make these choices – the other two main approaches within the deontological tradition are based on the idea of a social contract rather than individual acts. This is closer to discourse ethics which expects to operate at the level of the society or group. John Locke (1980, orig. 1689) based his approach on the idea of a set of natural human rights that society should enshrine, and this was influential in constructing the US Constitution and the UN's Universal Declaration of Human Rights. More recently, John Rawls (1971) introduced the idea of a 'veil of ignorance' in determining an appropriate set of social rules. Suppose that you knew nothing about your own personal characteristics (e.g. gender, race, disability) or position (e.g. wealth, class) in a society. If you were then asked to decide on the rules for that society, surely you would choose a set of rules that were equally fair to all so that you would not be disadvantaged whatever situation you found yourself in. This is quite an attractive idea, and there has been considerable debate between Habermas and Rawls (Habermas 1996). These social contract principles can be seen to underlie the idea of codes of practice for professional societies.

The third major strand of ethical theory, which dates back to Aristotle (2000) but has been developed in recent time by MacIntyre (1985), is known as virtue ethics or communitarianism. Aristotle was not concerned with the consequences of acts, or dutiful acts in themselves, but rather with a whole person and their way of behaving. He argued that people should develop emotions, personality and moral habits such that they 'naturally' behaved in a way that led to the well-being of the individual and the wider society. These ideas about what constitutes a virtuous and good life have been taken up by MacIntyre as a reaction to the ahistorical individualism assumed by the deontologists, especially Rawls. MacIntyre argues that we only become socialized as human beings through our development within a particular community and that we inevitably take on the codes and values of that community. This means that values and practices always remain relative to a particular community and there can be no external standpoint from which to judge them. While we can see that there certainly must be some truth in this argument, it leads to difficulties in arbitrating conflicts between cultural systems as are very much present in the world today (Habermas 2001).

10.3 Business ethics

For much of their development, business and corporations could argue that their only real commitments were to maximize profits and shareholder wealth within the confines of the law. However, the world has changed significantly during the last ten years and now there are few organizations that do not recognize that they have significant responsibility to local and global societies beyond simply making profit.

Many factors have led to this shift.

- Major corporations have been found not to be playing by the rules of the game, e.g. the Enron, Arthur Andersen and WorldCom scandals; human rights violations; and collaboration with repressive regimes (Palazzo and Scherer 2006; Sethi 2002).
- The effects of globalization means that some corporations are both economically and indeed politically more powerful than many nation-states (Beck 2000; Matten and Crane 2005). Even powerful world states such as the US have their policies shaped by corporate interests such as oil (re Kyoto) and defence. Moreover, when things go wrong, especially in the financial markets, it almost instantly damages the whole world economy.
- The rise of fundamentalism has also brought a much greater recognition of the importance of cultural and religious differences in values and behaviour

which cannot be simply effaced in the name of profit. The rise of ethical consumerism and investment has also demonstrated that companies have to take into account the ethical concerns of their consumers and indeed shareholders.

- Finally, and perhaps most significantly, the rather reluctant acceptance of the reality and consequences of global warming has led even hardened executives to accept that they are part of a problem that goes beyond short-term stock valuation or even long-term shareholder wealth.

All this means that there is now an increasing concern with the question of how corporations, and their employees, *ought to* behave, and this leads us to consider ethics as the appropriate theoretical and philosophical domain.

By way of reviews of this complex area I shall use Werhane and Freeman (1999), Garriga and Melé (2004) and Lee (2008). Garriga and Melé give an overview of different theories of corporate social responsibility distinguished in terms of their focus on economics, politics, social integration or ethics. Historically, we can see that initially there was a separation of ethics from business performance. Business's primary aim was economic performance and the maximization of shareholder (and executive) wealth (Friedman 1962) while social responsibility was voluntary and to some extent antithetical to business performance.[1] This stockholder or instrumentalist view has continued to underpin the more recent theories of competitive advantage (Porter 1985; Prahalad and Hammond 2002). Perhaps Bowen (1953) was the first to argue systematically that businesses, because of their great power and influence, were obliged to be socially responsible.

The next major phase was the development of theories of corporate agency – that is, conceptualizing corporations as morally responsible agents. There are various approaches that draw on different ethical traditions, for example Aristotelian (Solomon 1992), human rights (Matten and Crane 2005), and Rawlsian social contract (Donaldson and Dunfee 1995). We will illustrate the approach by looking at Donaldson and Dunfee (D&D) as an example. Their approach is aimed at overcoming one of the major problems of business ethics in the globalized world – how one reconciles differing cultural and religious practices. To what extent is it possible to generate genuinely universal norms?

D&D imagine that there will be some generally accepted social contract applying across the business world and that this in turn will allow for specific, micro-contracts in particular circumstances. This is because, D&D argue, in practice managers always have a bounded *moral* rationality. They cannot know fully the facts, or future consequences of their actions; they cannot have a perfect understanding of moral theory; and we do have to recognize legitimate differences in norms of practice across cultures, for example the giving of gifts. This means that the macro contract must allow for, and specify, a degree of *moral free space* or 'wiggle-room' within the micro-contracts. However, there must be limits to this; here D&D suggest the idea of *hypernorms*, norms that are genuinely universal and accepted by all. Their suggestions for hypernorms are basic human rights such as personal freedom, physical security and political participation, and the obligation to respect the dignity of every human being.

The second major approach to business ethics, and in fact to corporate strategy generally, is stakeholder theory. This involves recognizing that an organization depends for its successful operations on a range of different groups or stakeholders and therefore owes some duties to them. Two divisions within the field concern the reasons why stakeholders are important, and the range of stakeholder groups to be considered. For the first we can distinguish between the managerial or instrumental view and the normative view (Freeman 1999). The instrumental view is that stakeholders are important purely in terms of managing the company better (Donaldson and Preston 1995), while the normative view argues that companies *ought* to be concerned about their effects on various stakeholders for moral reasons (Evan and Freeman 1988). In the second debate the narrow view would include only those necessary for the survival of the corporation, whereas the wider view would include all groups that benefit from or are harmed by the activities of the organization. Theorists have drawn on a range of ethical positions including Kantianism (Bowie 1999), Rawlsianism (Phillips 2003), and extreme libertarianism (Freeman and Phillips 2002). There has also been a limited use of discourse ethics itself, which will be described below, but so far no employment of critical realist ethics.

10.4 Habermasian discourse ethics

I shall describe discourse ethics in three stages: the theory of communicative action from which it is derived; the initial formulation of discourse ethics; and then later developments into a theory of deliberative democracy. From this, its application within the business world can be debated.

10.4.1 Theory of communicative action

This will be a brief overview, as it is already well described elsewhere (Klein and Huynh 2004) and was discussed briefly in Section 8.4.3. The theory of communicative action (TCA) (Habermas 1984; Habermas 1987) argues that the most fundamental characteristic of human beings as a species is our ability to coordinate our actions through language and communication and, further, that the ability to communicate is grounded on the capacity to *understand* each other. Thus the primary function of communication is the construction of understanding and then agreement about shared activities. Humans do, of course, engage in other activity: for example purposive instrumental action in solving a problem or reaching a goal; or strategic action where communication is used to achieve personal ends through some form of deception or control. But even in this latter case, understanding is a necessary prior condition.

Habermas therefore sees communication oriented towards reaching agreement as the primary, and most common, form of communication, and proposes that the principle means of reaching agreement is through rational discussion and debate – the 'force of the better argument' – as opposed to the application of power, or the dogmas of tradition or religion. Habermas elucidates the nature of a 'rational' argument or discourse in terms of two concepts: (i) that contentions or utterances rest on particular *validity claims* that may be challenged and defended; and (ii) that the process of debate should aspire to being an *ideal speech situation.*

Whenever we actually say something, make an utterance, we are at least implicitly making claims that may be contentious. These validity claims are of three types,[2] and each one points to or refers to an aspect of the world, or rather analytically different worlds. These three types of validity claim are:

- *truth* – concerning facts or possible states of affairs about *the* material world;
- *rightness* – concerning valid norms of behaviour in *our* social world;
- *sincerity* (truthfulness) – concerning *my* personal world of feelings and intentions.

In our everyday discussions and debates, disagreements and misunderstandings develop and these lead to one or more of the validity claims being challenged. It is then up to the speaker to defend the claim(s) and possibly challenge the opponents. The discussion is now at a meta level compared to the original conversation. In order to achieve a valid, i.e. rational, outcome the discussion should occur in such a way that it is the arguments themselves that win the day rather than distorting aspects of the people involved or the social/political situation. Such an ideal speech situation (which can only ever be a regulative ideal to aim at) should ensure (Habermas 1990, p. 86):

- All potential speakers are allowed equal participation in a discourse.
- Everyone is allowed to:
  - question any claims or assertions made by anyone;
  - introduce any assertion or claim into the discourse;
  - express their own attitudes, desires or needs.
- No one should be prevented by internal or external, overt or covert coercion from exercising the above rights.

Habermas argues that these are not merely conventions, but inescapable presuppositions of rational argument itself. Thus someone engaging in an argument without accepting the above is either behaving strategically (deception) or is committing a performative contradiction (hypocrisy).

10.4.2 Discourse ethics

Discourse ethics (DE), which is somewhat badly named as we will see, stems almost directly from the TCA through considering actions in general rather than just communications. It is clearly Kantian in thrust, although with a very significant re-orientation, but also sweeps in to some extent utilitarian and communitarian concerns.

Beginning with the traditional ethical question 'how should we act?', Habermas (1993b) recognizes that such questions occur in different contexts. We may begin with basic *pragmatic* or purposive questions about the best ways to achieve particular ends. How to earn some money? How to fix the car? These often concern problems in the material world and they may be quite complex. Their resolution may well require information, expertise and resources. Many of the problems that occur within a business context are often seen like this, and in that domain they would be classed as 'hard' rather than 'soft'. In terms of ethical theory this relates to the consequentialist approach, in which actions are judged in terms of their effects and consequences but only in the self-interests of the actor(s) concerned.

The question might, however, be rather deeper. What if the goals or ends to be achieved are themselves in question, or if the means to be used raise ethical or moral issues? Here we are concerned with the core values and the self-understanding of a person or a community. What kind of person am I, or what kind of group are we, that we should have these particular values and behaviours? These questions concern what Taylor (1989) called strong preferences, to do with our being and way of life, rather than simply weak preferences such as tastes in food and clothes. Habermas calls these types of questions *ethical* questions, in contrast to *pragmatic* questions discussed above and *moral* questions discussed below.

Within the pragmatic domain, efficacy is the test – does the action work? Does it have the desired effect? But within the ethical domain, goodness or virtue is at issue. Does the action accord with and develop the actor's own existential identity and self-understanding? This clearly picks up on the Aristotelian and communitarian positions that emphasize the importance of developing the good life within one's community. Although the pragmatic and the ethical have very different concerns – the efficacious and the good – they are similar in that they are both oriented towards the self-interests of particular individuals or groups – the question is, what is efficacious or good *for us*? It is when one goes beyond that perspective to consider what might be good *for all* that one moves into the domain of *moral* questions. And this is really the focus of discourse ethics.[3]

> We should not expect a generally valid answer when we ask what is good for me, or good for us, or good for them; we must rather ask: what is equally good for all? This 'moral point of view' constitutes a sharp but narrow spotlight, which selects from the mass of evaluative questions those action-related conflicts which can be resolved with reference to a generalizable interest; these are questions of justice.
>
> (Habermas 1992a, p. 248)

So, while discourse itself applies to all three domains, the main thrust of discourse ethics is actually moral questions, that is, those that concern *justice for all*; those that transcend the interests of any particular individual, group, nation, or culture but that should apply equally for all people. Habermas's approach is clearly Kantian in that he is interested in that which is universalizable, but he effects a major transition away from the subjective thoughts or will of the individual agent (a monological focus) towards a *process of argumentation and debate* between actually existing people (a dialogical focus). This marks DE out from other approaches, as Habermas does not see this as just an analytical procedure or thought experiment; he intends that such debates, especially within society as a whole, should actually occur. We can see now how discourse ethics is intimately related to TCA: the three domains, the pragmatic, the ethical and the moral correspond with the three worlds; and the whole approach is embedded within the processes of communicative action.

How should we judge whether an action-norm is universalizable? Kant's categorical imperative is an exercise conducted from a particular person's viewpoint: what do they think would be suitable for all? We need to go beyond that and test whether such a maxim or norm can also be accepted by all of those affected by it. This leads to a reformulation of the categorical imperative in what Habermas calls the *discourse principle* (D): 'Only those norms can claim to be valid that meet (or could meet) with the approval of all affected in their capacity as participants in a practical discourse' (Habermas 1992b, p. 66). This is a general statement about what would constitute a valid norm and has two essential parts: that the norm must be agreed or approved by all those affected, and that this must occur through an actual process of discourse. This is analogous to the truth of descriptive statements (Habermas 1999a). A statement is *true* if what it claims about the world is in fact the case. This is a definition but it does not tell us how to find true statements. Equally, a moral is *right* if all affected have participated in a fair discussion and agreed to it.[4] But D does not specify what such norms might be, nor what might be the process of discourse. The latter point is developed through a further *universalization principle* (U) which outlines how such norms might be arrived at: 'A norm is valid when the foreseeable consequences and side-effects of its general observance for the interests and value-orientations of each individual could be jointly accepted by all concerned without coercion' (Habermas 1999a, p. 42.)[5] The point of this process is to try to generate a *common will* and not just an accommodation of interests. That is, the participants should become convinced that it is genuinely the best way for all of them to resolve their common differences. To this end, (i) the mention of 'interests' and 'value-orientations' refers to the participants concerns within the pragmatic and ethical domains respectively. (ii) Participants should try and genuinely take on the perspectives and roles of the other, and be prepared to modify their own. (iii) Agreement should be based, as always, on force of argument rather than force of power.

10.4.3 Towards deliberative democracy

Habermas has always had as one of his primary concerns politics and the nature of the state. In the 1960s he argued against increasing instrumentality and technocracy in *Towards a Rational Society* (Habermas 1971), and in the 1970s he analysed the developing crisis in Western societies in *Legitimation Crisis* (Habermas 1976). During the 1990s he developed his communicative and moral theories into a powerful model of the nature of democratic society within the postnational and multicultural age (Habermas 1996; Habermas 1999b; Habermas 2001). This has generated considerable debate within political and legal circles (Dryzek 2002; O'Flynn 2006; Parkinson 2006).

Societies are governed by laws, and laws embody, in part, norms of expected behaviour. There is, therefore, an intimate connection between morality, with its concern for rightness and justice for all, and the law and its need for legitimacy. The law also ultimately rests on the discourse principle (D), which defines valid norms, but there are significant differences between morality and law. Morality, as we have seen, is a domain drawn narrowly to include only those norms that can gain *universal* acceptance; it thereby excludes the ethical domain of individual or community values and conceptions of the good, and the pragmatic domain of goals and self-interest. The law cannot do that, however. It must operate in the real world and be able to regulate all three domains together. Moreover, and perhaps partly because of this, the law is positive as well as normative: it can take action and apply coercion and sanctions as well as claiming validity, whereas the moral domain rests on individuals and their consciences for its enactment.

These relations are illustrated in Figure 10.1 (Habermas 1996). At the top is the discourse principle which then splits into two – the moral principle and the

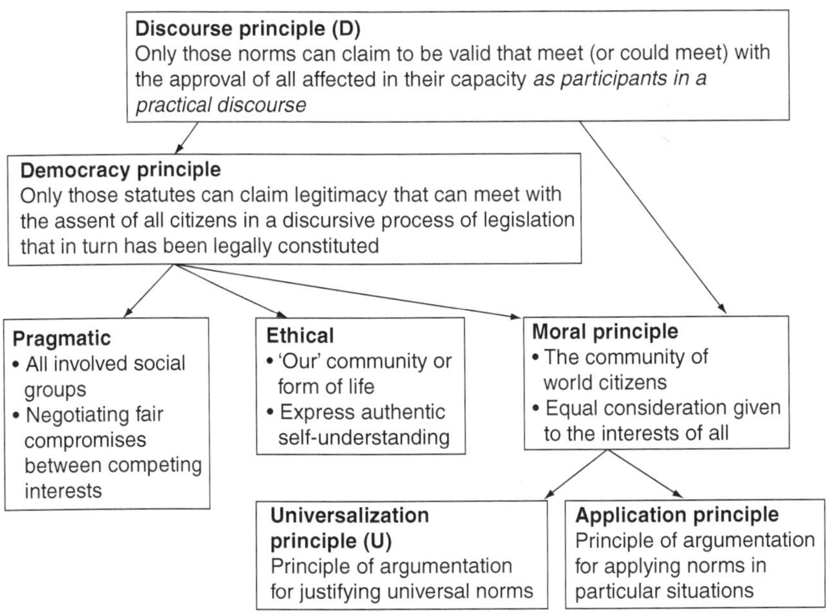

Figure 10.1 Varieties of discourse.

democracy principle – although as can be seen these are at different levels. The democracy principle governs those norms that can be legally embodied and gain the assent of all citizens through a legally constituted legislative process. Such laws have to deal with questions that arise in all three domains – the pragmatic, the ethical, and the moral. Each domain involves different reference groups and different discursive procedures. Moral questions are governed by considerations of fairness for all and ultimately relate to the world community. Moral norms can be justified through the universalization principle (U), but there also needs to be discourse about the application of norms to particular situations, the application principle. Ethical questions concern issues of self-understanding of particular communities or forms of life and are highly relevant to the multicultural societies that exist nowadays. Pragmatic questions involve bargaining and negotiating fair compromises between competing interests.

Morality and the law are thus distinct but complementary. Morality is a domain where people agree to take on duties and particular forms of behaviour because they reach consensus through debate that the norms are universally applicable. The law should enshrine these norms but will also have to include many more specific norms to deal with ethical conflicts between different communities and pragmatic conflicts between different interests. Habermas envisages stages through which such debates may occur (Habermas 1996, p. 164). Initially, proposals or programmes for action are brought forward and these are evaluated in generally technical terms, based on information, knowledge and technical expertise, an example of the classic decisionistic approach of evaluating different means for accepted ends. Often, however, the ends, that is the values and interests themselves, are seen to compete; discourse now needs to change to another level. There are now three possibilities. First, the issues may involve *moral* questions: that is, questions that need to be solved in the interests of all, for example social policies such as tax, health provision or education. Second, the issues could involve *ethical* questions that may differ between different communities and may not be generalizable, such as immigration policies, abortion, or the treatment of the environment and animals.

Or, third, the problem may not be resolved either through general assent or the strength of a particular value because of the range of different communities and interests involved. In these cases one has to turn to bargaining rather than discourse. The parties involved need to come to a negotiated agreement or accommodation rather than attain a consensus. This is *not* a rational discourse (in Habermas's terms), since the parties involved will be acting strategically and may well employ power, and because the parties may agree for different reasons, whereas with a moral consensus the parties will agree for the same reasons. Nevertheless, rationality and the discourse principle can be applied to the *process* of negotiation if not its actual content.

So deliberative democracy can be seen to weave together a whole variety of different forms of discourse and communication, involving rational choice and the balancing of interests; ethical debates about forms of community; moral discussion of a just society; and political and legal argumentation. This complexity occurs not just in the traditional institutions of politics and the law, but increasingly in what Habermas refers to as the voluntary associations of civil society (Habermas 1996, Chapter 8). The whole third sector of community and voluntary groups, pressure groups, NGOs, trade associations and lobbyists, underpinned by the explosion of communication technologies, now occupies the space between the everyday communicative lifeworld, the economy and the state. Those in the sector sense and respond to issues and concerns that arise within the public sphere and channel them into the sluice-gates of the politico-legal centre.

10.5 Ethics in critical realism

Bhaskar's critical realism (CR) has been developing as philosophy of science and social science for many years, as we have seen. The ethical theory, although implicit and sometimes explicit in Bhaskar's earlier work, has become more developed in dialectical critical realism (DCR) (Bhaskar 1993; Bhaskar 1994). I shall not cover Bhaskar's work beyond DCR.

CR's view of morality has two main principles:

- **Moral realism**: that is, that there are moral truths in the intransitive domain independent of the subjective views of individuals or traditions, ultimately grounded in characteristics of human nature.
- **Ethical naturalism**, which implies that we can, through social science, discover what these moral truths are. This involves moving from facts, about the way things actually are, to values, i.e. how they should be, and thus requires a refutation of Hume's law that you cannot derive 'ought' from 'is'. This is done by way of the concept of 'explanatory critique'.

The ethical approach can be expressed in terms of four stages – the *ethical tetrapolity* – which can be set in motion within a variety of different contexts: speech or discourse, action, existing morality, or social science itself. Within DCR these dynamics are all termed dialectics, as in the dialectic of discourse, the dialectic of action or the dialectic of morality. I will explain the tetrapolity within the realm of discourse, as that is most easily comparable with discourse ethics, before evaluating the extent to which I think the arguments succeed.

The ethical approach rests on four main arguments which follow on from each other:

A That social science is evaluative, not value-free (Bhaskar 1993, Chapter 3.7)
Traditionally, science has rested on the Humean premise that facts and values are separable, and science is only concerned with facts; and, *a fortiori*, that you cannot logically derive an 'ought' from an 'is'. The first argument of the ethical approach establishes that (social) science in not value-free but unavoidably evaluative.

The subject matter of social science, the phenomena of the social world, is itself intrinsically value-laden, and it is wrong for social science to try and avoid this by redescribing the phenomena in neutral terms. For example, while (a) 'X was murdered' and (b) 'X ceased breathing' may both be true descriptions of the same event, (a) is to be preferred because: (i) it is more accurate and particular – (a) implies (b) but not vice versa; (ii) b) tends to carry the presumption that X died naturally, since that is more common, when that is not in fact the case; and (iii) (a) maximizes the explanatory power of the theory required to explain it. To explain death requires only biology; to explain murder requires sociology and psychology as well.

Thus, social science is inevitably and properly evaluative.

B That one can derive 'ought' from 'is' (explanatory critique) (Bhaskar 1993, Chapter 3.7)

The next stage is to go beyond simply being evaluative to deriving normative implications – i.e. guides for action.

It is in the nature of social science to study social beliefs (that is, the actual beliefs people hold about society) and to be able to judge their truth or falsity. It is also possible to show that there are structures within society that generate and maintain both true and false beliefs. So:

1 Where science can demonstrate that a widely held belief is false, and
2 identify structure(s) that maintains the false belief, and
3 identify actions that would displace the structure(s), then
4 (*ceteris paribus*) it can negatively evaluate the structure(s) and
5 positively evaluate the actions to remove them.

The same basic argument can be applied to social conditions that are considered immoral rather than false beliefs, i.e. conditions that obstruct the realization of freedom: for example, unnecessary constraints and unwanted ills. The *ceteris paribus* clause will be discussed below.

C That moral judgments imply commitments to action (theory-practice consistency) (Bhaskar 1994, Chapter 7.1)

So far the arguments have been at the level of social science, but what about commitments of the individual towards taking action? The argument can be put in terms of discourse (speech acts) or agency more generally.

1 Where one expresses a judgement of a moral or ethical kind (*expressively veracious*) there is an implication of axiological commitment, that is, solidarity with the addressee to remove unwanted constraints or unnecessary ills. In other words, where someone criticizes some aspect of society as being morally wrong, there should be a commitment to changing it, if possible.
2 The speech act should be taken as trustworthy by the addressee, implying that they (and the addressor) should act on it (*fiduciariness*).
3 This leads to the need for *explanatory critique* (theory) to understand the reasons for the situation, and

4 *emancipatory axiology* (practice) to take action to remove them,
5 contributing to concrete universalized freedom of all.

D That these may be universalized (Bhaskar 1994, Chapter 5.3, Chapter 7.1)
The final step is to go from addressing a particular problem or constraint to a commitment to address all such constraints.

Once a commitment has been made, through a fiduciary[6] remark, to remove a particular ill or constraint, the addressor and the addressee are logically committed to removing other similar ills and constraints and, similarly, are committed to removing constraints and ills as such, and ultimately to changing the society that maintains them. 'So the goal of universal human flourishing is implicit in every practical deed and every fiduciary remark' (Bhaskar 1994, p. 148).

10.5.1 Assessing critical realist ethics

For me, there are two major strengths of CR ethics – the idea (which it shares with other critical traditions) of the necessity of emancipatory critique; and some of the specific arguments that Bhaskar makes concerning the relations between facts and values. But there are many problems in ever realizing it in practice.

Most traditional ethical theories – deontological, utilitarian or contractarian – generally take a subjectivist, individualist position. They specify in some way or another how individuals should act and then assume that all will be well so long as they make the right decisions and then act on them. In contrast, CR, following in the line of Marxism and critical theory, recognizes that people are not transparent to themselves, holding many unacknowledged and potentially false beliefs, and are constrained by structures and mechanisms within society. Thus we cannot expect that people will simply think and do the right things; we need social science that is enlightening in revealing false beliefs and empowering in generating alternatives. In this respect, critical realism is an advance through its more sophisticated philosophy of science and social science, and its model of human activity (the TMSA discussed in Chapter 6).

In terms of the four arguments above, I do think that arguments A, B and C are powerful in establishing at least the *principle* that social science is intrinsically and unavoidably evaluative, and thereby committed, *in principle*, to critique of the status quo. However, as we shall see, going further to practical applications of these principles is highly problematic. Several authors have put forward criticisms of critical realist ethics (Hammersley 2002; Hostettler and Norrie 2003; Lacey 1997; Sayer 2000), and I shall summarize them together with some of my own.

The first set of criticisms concern primarily argument B. This says, simply, that social science can determine that particular social beliefs are wrong; that there are mechanisms sustaining these beliefs; and that there could be preferred alternatives that should therefore be realized. The main arguments against it are that it involves a simplistic view of both science and the complexity and openness of modern society (in contrast to most of the rest of CR).

- The fallibilist nature of science (accepted by CR) means that we can never know for sure that particular beliefs are actually wrong; that particular structures sustain them; or what the actual effects of alternatives might be. This is exacerbated by the fragmented nature of social science itself with significant debates about philosophies, basic concepts (e.g. 'class'), and degrees of applicability.
- The complex and open nature of society (again accepted by CR) means that there is unlikely to be a simple one-to-one relation between a mechanism and a resulting belief or ill. Rather, a range of mechanisms interact with each other in complex ways generating a range of actualities some of which may be judged to be beneficial as well as harmful. So removing a mechanism, or part of a complex structure, may have unforeseen and unwelcome effects. Equally, it is difficult to construct alternatives that can be seen unambiguously as both desirable and feasible. These problems are generally hidden within the *ceteris paribus* clause that Bhaskar uses.
- These problems are exemplified by the lack of any practical examples of emancipatory critique within social science.

The second criticism concerns especially arguments C and D. The crux of the argument here is that once one accepts that a particular state of affairs is wrong, one is committed to taking action to change it for fear of committing theory-practice inconsistency. I think that it is reasonable that if one believes something is wrong, one should not take action which knowingly supports it, although even that is difficult because of the unacknowledged conditions and unknown consequences of our actions. But that does not commit one to taking direct action against it. After all, we all live in a world with many horrendous events that we would wish to be different but we cannot then be expected to devote our lives to trying to change them all, or indeed any of them specifically. We have to live as best we can in an imperfect world, and have to try to be consistent in aligning our activities with our values, but even that is very difficult given the complexity and uncertainty described above.

The third set of criticisms concern the assumption (developed more strongly in *From East to West*, Bhaskar 2000) that the primary problem is constraints and ills forced on people by society and that if only these were removed people would share a universal set of values. However, this is very much an assumption and one that there is little evidence for in the world at the moment. Rather, we experience a world that is strongly divided, not only in terms of pure interests, the haves and have-nots, but perhaps more substantively in terms of culture, ethnicity and religion. This means that people, already enmeshed in a moralized world, will come at issues with fundamentally different views. For instance, with regard to argument A, one side in a conflict might describe the killer as a 'freedom fighter' while the other side would talk of a 'terrorist'. From what standpoint could a critical realist judge one to be right and the other wrong – only from another, equally value-laden position? Through Western eyes, Muslim treatment of women might be seen as unnecessary and unwanted constraints, yet for a Muslim, Western behaviour may be equally offensive.

These deeply entrenched value positions affect not only the starting point of the arguments in terms of problems to be addressed, but also recur at each stage in terms of structures and mechanisms that may be generating the situation and possible alternatives to them. They are not really addressed at all by Bhaskar, except for a brief nod towards communitarians (Bhaskar 1994, p. 160) where he accepts both that moral beliefs are diverse and that moral truths are relative because of the open nature of society. This latter point seems to undermine much of his previous argument, since if moral *truth* (as opposed to *knowledge*) is relative to time and culture how can we privilege one over another? I suggest, as does Sayer (2000, p. 167), that this perhaps leads into Habermas's discourse ethics.

10.6 Comparing Habermas and Bhaskar

There are clearly many *prima facie* resonances between the two approaches in general, and Bhaskar mentions Habermas's work on several occasions, sometimes to point out differences (Bhaskar 1994, p. 142) and sometimes to talk of a rapprochement (Bhaskar 1994, p. 160).

10.6.1 Similarities between Habermas and Bhaskar

Critical social science

First, the fundamental ideas that social science is inevitably evaluative and that it should play the role of explanatory critique were both explicit in Habermas's early work – see for example *Towards a Rational Society* (Habermas 1971) and *Theory and Practice* (Habermas 1974). That was, in many ways, the whole purpose of a critical science. Also, both ideas use essentially the same form of argument, what Bhaskar calls transcendental or retroductive and Habermas calls rational reconstruction. This involves taking some generally agreed phenomenon (e.g. experimental activity or human communication) and asking what the world must be like for this phenomenon to occur as it does.

What is interesting is that in his more recent work, certainly since discourse ethics, Habermas makes virtually no mention of critical theory or critical social science. I am not sure whether that is because it is just taken for granted as a background to the whole project, or whether it signals that Habermas is no longer committed to such a view. Certainly he seems much more pessimistic now. In 'What Theories Can Accomplish – and What They Can't' (Habermas 1994, p. 116) he writes:

> All social theories are highly abstract today. At best, they can make us more sensitive to the ambivalences of development: they can contribute to our ability to understand the coming uncertainties … they can open our eyes to dilemmas that we can't avoid and for which we have to prepare ourselves.

While in a chapter called 'The Relationship between Theory and Practice Revisited' (Habermas 2003b, p. 285) he admits that 'Philosophy thus no longer positions itself as a pretentious countervailing power against the entire modern world'.

Discourse

Both theories can be set within the framework of discourse and its presuppositions. Habermas is wholly oriented towards discourse or communicative action, while Bhaskar's theory is realized in several domains. With regard to discourse, however, both see it as a fundamental form of human activity that brings with it certain claims or assumptions. For Habermas, implicit in every (non-strategic) speech act is the idea of reaching understanding through unfettered debate; for Bhaskar, implicit in every fiduciary remark is a commitment to an emancipated society.

Moreover, there are interesting linkages between Bhaskar's fourfold judgement form and Habermas's validity claims. For Bhaskar, when a person makes a judgement which is *expressively veracious* (roughly, a claim that expresses, very well, how things really are) the claim must be *imperatival-fiduciary* (trust me, you can act on it); *evidential* (there must be good grounds for believing it) and *descriptively accurate* (this does represent how things are). This is related to Bhaskar's four degrees of truth (the truth tetrapolity): the weakest level is *fiduciary*, 'just trust me'; the second level is *epistemological*, warranted assertability – there must be evidence or grounds for it; the third level is *expressive-referential* such that the proposition does describe how things actually are; and the fourth is *ontological* or *alethic* – the existence and causal grounding of things in themselves.[7]

In terms of validity claims, sincerity clearly relates to the fiduciary in concerning the trustworthiness of the speaker. Truth would seem to relate to expressive-referential, as it concerns claims as to states of affairs in the material world at least, but the rightness of norms does not fit well. This is because of Habermas's anti-naturalism and consensus theory of truth, which will be discussed below.

The ideal society

Both writers express very similar views about the nature of an ideal society towards which we should try to move. Bhaskar uses the Greek term *eudaimonia* to describe a happy and flourishing society in which everyone is free from unnecessary constraints on their freedom. This recognizes that people should be free to be different to the extent that this does not restrict the freedom of others. The freedom of each is a necessary condition for the freedom of all. Habermas is less specific about some idealized society as his approach is more procedural, concerned with specifying rational procedures for communication and discourse that would then allow participants to generate their own moral norms, but he does say (Habermas 1999a, p. 40):

> moral concern is owed equally to persons both as irreplaceable individuals and as members of the community … equal treatment means equal treatment of unequals who are nonetheless aware of their interdependence.… The equal respect for everyone else demanded by a moral universalism sensitive to difference thus takes the form of a nonleveling and nonappropriating inclusion of the other in his otherness.

However, they differ in terms of where society is coming from and how it is to get there. For Bhaskar, the problem is largely the oppressive power relations of existing societies and the *modus operandi* is action, if necessary direct action, to remove these constraints. For Habermas the concern is more how, in today's globalized society, conflicting collectivities, be they based on religion, culture or ethnicity, can reconcile their differences in a way that is satisfactory for all. And the answer is debate and discourse in which participants genuinely try to take on the perspective of the other (Habermas 2003b, p. 235): 'For given a pluralism of legitimate world views, conflicts of justice can be resolved only if the disputing parties agree to create an inclusive We-perspective by mutual perspective-taking.'

Other commonalities

I have discussed above how Habermas's theories have led to the development of sophisticated ideas concerning the nature and role of law in modern societies, and also the concept of deliberative democracy and the importance of third-sector organizations such as NGOs, pressure groups and so on within that. Bhaskar too sees an important role for participative democracy or participation-in-democracy (not perhaps quite the same as deliberative democracy but certainly close) and recognizes the role of a variety of organizations within this. He also accepts that this may have to be representative rather than fully inclusive or participative (Bhaskar 1994, p. 156).

Second, both embrace forms of universalization as a foundation for morality, although in different ways. For Habermas, universalization is specifically in terms of norms that apply to all people, at least all those who are affected by something, and this is what distinguishes moral from ethical questions. It is different from the Kantian categorical imperative in being the result of a particular discursive process rather than abstract and general. For Bhaskar, universalization concerns the extent to which judgements should apply in other, similar circumstances. If the judgement is true and truthful then the same reasons and results should apply and the speaker should be willing to affirm it in another person's situation. It is thus a test of consistency (sincerity) and truth (replicability) (Bhaskar 1994, p. 65). Bhaskar emphasizes that this applies to the concrete individual with their particular circumstances, rather than the generalized other. However, this raises questions about the extent to which any situation, or person, is similar to another – are they not all at some level unique? Habermas addresses this question in terms of a discourse of applicability which considers whether particular norms are applicable in individual circumstances.

Finally, both claim that their approaches can include other ethical theories such as communitarianism, virtue ethics, deontology, and so on.

10.6.2 Differences between Habermas and Bhaskar

For all that there are similarities in priorities and approach, there are at least two major and related divergences – the reality of moral truths and the role of (social) science in discovering them and thus in bringing about the eudaimonic society.

For Bhaskar, moral truths are real and they may be discovered through science. He maintains a strong distinction between moral principles as they actually exist at the moment, which may be distorted and false; and moral truths, which can be generated through explanatory critique. Habermas, on the other hand, maintains that valid moral norms (he would not call them moral *truths*) are *constructed* by people coming to agreement through a process of discourse. Thus, whereas Bhaskar claims an outside standpoint from which to critique existing views, Habermas holds that there is nothing other than the result of a practical discourse to determine what moral norms should be. These would seem to be incompatible positions: either moral truths exist over and against existing peoples' beliefs, or they are actually determined by peoples' beliefs, but surely not both? I want to argue now that in fact that their positions are much closer than this stark contrast would suggest.

If we begin with Habermas, for most of his career he had a decidedly anti-realist and anti-naturalist stance. There were clearly different domains of knowledge which had their own appropriate methodologies and certainly there was a dislocation between the empirical sciences and the social sciences (Habermas 1978). One thing the sciences shared, however, was a consensus theory of truth: i.e. even in the empirical sciences, truth, following Peirce's pragmatism, was defined as that which was agreed by the scientific community through unfettered discourse rather than that which resulted from interaction with an independent reality.

However, as we saw in Section 8.4.3, in more recent work Habermas (2003b, p. 8) has made a significant shift towards what he calls an ontological rather than a purely epistemic conception of truth, and a 'weak naturalism' which puts him much closer to critical realism in general.

However, Habermas still draws a distinction between propositional truth and moral rightness. Claims to rightness are akin to, or analogous with, the concept of truth, but are not identical to it. Truth is discursively arrived at, in that what is taken to be true at a particular time is the result of debate and agreement, but it nevertheless has an outside referent that can demonstrate it to be wrong. Both truth and rightness are discovered in the same way, through discourse; and both are justified in the same way, warranted assertability; but a true proposition refers to an objective world whereas a right norm does not refer to anything outside the discourse. It is no more than the warranted assertability that those involved in the discourse have agreed that it is indeed worthy to be a universal norm (Habermas 2003a).

But this does not mean that rightness has no sense of externality at all, that it is purely a free construction. It can to some extent meet the two characteristics of objectivity – that it is not just up to us, and that it is the same for all of us. First, rightness does have to be *discovered* in the same way as truth. It is only after an actual discursive episode, in which social players with differing interests and values have had to battle to an agreement, that moral norms are established. They are created *through* the discourse; they cannot be determined by outside observers. Second, agreed norms may later turn out to be wrong, either because the premises or information available was limited or incorrect, or indeed because the open social world changes and develops and the situation becomes different. This is very much the case with environmentalism today – the current debates and discussions have been triggered by new knowledge. Third, there is the idea of universality – only those norms are right that can be agreed by all affected, and then enacted in bringing about a moral society (Habermas 2003a, p. 261):

> Insofar as we test the rightness of moral statements from such a universalist point of view, the reference point of an ideally projected social world of legitimately ordered interpersonal relationships can serve as an equivalent for the absent constraints of an objective world.

The second major difference between Bhaskar and Habermas, the role of science within explanatory critique, was discussed briefly above. Certainly Habermas now makes very little of the role of or need for science in moral argumentation. It is as if he no longer recognizes that participants in a debate may indeed be misinformed or have false beliefs which would thus render the results invalid. However, since the whole point of an ideal process of argumentation is that *all* information and knowledge are brought in, and that *all* views are open to critique, one might expect that scientific knowledge would figure strongly and would potentially lead to changes of view and attitude in support of the 'better argument'.

But Habermas would see the difficulty in a moral debate as different from that of false belief. In everyday discourse, or even in scientific discourse, the primary concerns are matters of fact, and perhaps beliefs about their causes, but a moral discourse is driven by differences in (perhaps deeply held) convictions, values and interests. The requirement on participants is that they must be genuinely honest with themselves, and that they are willing to place equal weight on the views and values of other participants in order to surpass their differences and find a just outcome which they will all, as participants, have to live with.

We could perhaps summarize this in the following way. For critical realists, Habermas is still not realist enough even though he has moved in this direction. The realities of oppression and suppression still force themselves upon us in ways that go beyond merely moral debate. Critical realists would insist on the ontological reality of social structures and their effects, and would insist on the necessity for an emancipatory social science to go beyond the everyday assumptions of existing moralities. This would seem to place critical realism above or beyond discourse ethics.

Yet, as we saw in the criticisms of CR, is CR not too simplistic in its rather one-dimensional and reductionist views of causation and change? Do we not live in an extremely complex, interdependent and rapidly changing world in which social science is hard-pressed to explain what is happening at the time let alone come up with equitable and robust alternatives? At the same time, is it not the case that in fact many people understand only too well the realities of the world and the strong differences in value orientations and interests that are in play at this time? This would lead critical realism, with its fallibilist view of knowledge, to have to accept that in practice it is actually the agreements and commitments of participants within a discourse process, whether they are scientists/experts or ordinary people, which is necessary for development and change. Put this way, critical realism, despite its realism, has to come to depend on discourse to decide what it takes to be true at any particular time and thus may ultimately depend on discourse ethics. Even Bhaskar recognizes that there will always be difficult decisions to be taken, even in utopia (Bhaskar 1993, p. 295):

> Such a [eudaimonistic] society would be an open process.... Contradictions would exist, of necessity. Difficult decisions would have to be taken, democratically – at a plurality of spatial and organizational levels and spheres of interest – by sometimes circuitous decision-making routes. There would be competing conceptions of the details of the eudaimonistic society, grounded in competing theories of four-planar social being, almost inevitably represented by competing parties.

How are these difficult decisions to be taken and competing interests reconciled if not through discourse and debate?

We can therefore argue that both positions complement and contribute to each other, although there does remain a small but significant fissure between their respective positions on social reality. This could perhaps be bridged if Habermas made a further move in this direction to recognize the objectivity of at least some aspects of society.

10.7 Applying discourse ethics and critical realism in management

There have been some applications of DE already in management, but none that I can find of critical realist ethics. DE has been advocated in two main ways: concerning the role of corporations as a whole within society, drawing on the later theory of deliberative democracy; and also at the level of communications within organizations.

Reed (1999a; 1999b) has used DE as the basis of a normative stakeholder theory of the firm, arguing that the distinctions between legitimacy, morality and ethicality provide a more sophisticated and comprehensive approach to dealing with the normative bases of stakeholder claims. He also argues that the underlying communicative theory goes beyond the abstract notions of a Rawlsian veil of ignorance towards actual debate and discourse, and a recognition of the realities of compromise and bargaining. Smith (2004), in part developing from Reed's work, argues that increasingly companies will not be able to achieve their long-term strategic aims by acting in a purely instrumental, pragmatic manner – they need to become engaged within the moral and communicative spheres of society as a whole. In a similar vein, Palazzo and Scherer (Palazzo and Scherer 2006; Scherer and Palazzo 2007) argue that corporations need to become politicized in the sense that they need to become genuinely political agents within an increasing globalized, 'postnational' (Habermas 2001) world. 'These phenomena need to be embedded in a new concept of the business firm as an economic *and* a political actor in market societies' (Scherer and Palazzo 2007, p. 1115, original emphasis).

Moving to communicative action as such, Meisenbach (2006) has attempted to operationalize Habermas's universalization principle (U) to guide those conversations within an organization that have a moral dimension, i.e. that potentially affect all those within the community, and I will take up her proposals below. DE has also been suggested as a basis for theorizing moral principles in decision-making in organizations (Beschorner 2006; de Graaf 2006) and as a basis for ethical auditing (Garcia-Marzá 2005).

For myself, I suggest that there are several general contributions from DE, and to some extent from critical realism. First is the idea of *practical discourse*. DE is unlike *all* other ethical theories in that it requires actual discussion and debate among real people who may be affected by a norm or proposal, and it accepts the outcome as that which is morally correct, assuming of course that the debate was sound. In this, it would seem to have the *potential* for bringing about on-going, practical resolutions of moral and ethical concerns.

The second contribution is its emphasis on *universalization*. DE distinguishes moral issues that concern everyone involved in a particular situation from ethical and pragmatic ones which are relative to particular individuals or groups. It therefore pushes us to consider, *and involve*, as wide a range of stakeholders as possible in decisions and system designs. This is especially of concern today in a world with such fractured and antagonistic worldviews where involving all parties in trying to find shared ways forward seems the only possible strategy.

Third, DE is both more comprehensive, and in a particular sense more practical, than other ethical theories in recognizing that in the real world there are different types of issues, and different perspectives from which to approach them. As well as questions of justice, DE incorporates, to some extent, the concerns of utilitarianists and consequentialists in accepting pragmatic questions that need to be settled through bargaining and even the exercise of strategic action. It also recognizes the concerns of communitarians in accepting that some questions may well *not* generate universal, but only local, agreement and yet can still be the subject of rational discourse. Business and management, like law, also has to deal with issues in all three domains since, in the long term, *effectiveness* also requires an acknowledgement of the good and the just as well as the practical.

Moving to critical realism, its main contribution would seem be to management studies as a discipline and a pedagogy. There has long been a debate about the relationship between the discipline and its object of study – management practice. Traditionally, management studies was seen in either positivistic terms, generating generalized, abstract and value-free knowledge based on measurement and experimentation; and/or in functionalist terms, concerned with making management more effective on behalf of shareholders. This was generally called 'Mode 1' knowledge (Gibbons *et al.* 1994), as opposed to Mode 2 knowledge, which was more engaged, context-driven and problem-oriented. In either case, however, significant problems emerged in terms of what became known as the 'rigour vs. relevance' debate. The more rigorous, in positivist terms, that knowledge was, the less relevant and useful it tended to be; while the more practical and relevant, the less it was valued as knowledge as opposed to mere 'consultancy'.

But there is a third perspective, much more aligned to critical realism, known as critical management studies (Alvesson and Willmott 1992; Grey *et al.* 1996) that recognizes a greater degree of ambiguity between management theory and practice. On the one hand, management as a discipline needs to be able to hold itself away from the actuality of practice in order to be able to analyse and critique it. On the other hand, especially in terms of management education, it is the management discipline that is training the next generations of managers and so must be responsible for equipping them with more than simply functional techniques. Here, critical realism can play a major role (Syed *et al.* 2009) in demonstrating the value-laden nature of social science and providing secure philosophical underpinnings for an emancipatory study of management.

If these are the strengths of discourse ethics and critical realism, it has to be accepted that, as it stands, they are too abstract and idealized to be directly or practically utilized within business. So we need to consider to what extent they can be pragmatized without becoming entirely emasculated. Ways of doing this, linking into practical methodologies for problem solution and resolution, have been investigated in the domain of information systems (Mingers and Walsham 2008) and management science/operational research (Mingers 2011).

10.8 Conclusions

We live in a world in which businesses and corporations often have more power than nation-states and in which globalization has brought to the fore the deep divisions between cultures and religions. This makes it vital that managers consider the ethical and moral dimensions of their decisions, not only the economic ones. A range of traditional ethical theories have been deployed in business, but they often remain abstract and somewhat arbitrary in their application.

This chapter has considered two more modern ethical approaches – discourse ethics and critical realism – to see whether they have anything of substance to offer. They share a considerable range of commonalities while at the same time differing in their view as to the reality of moral truths. But, I argue, in many ways they complement each other and can potentially be combined together in fruitful ways. Even if this were done, as it stands they remain at too abstract and idealistic a level to be of direct practical use and research would be needed to pragmatize them without losing their essentially critical core.

Notes

1. There were interesting exceptions to this, such as Unilever and Cadbury's who historically integrated ethics with business.
2. There is a fourth – comprehensibility – concerning the understandability of the utterance itself.
3. Indeed, Habermas accepts that it should really have been called 'a discourse theory of morality' rather than ethics (Habermas 1993a, p. vii).
4. For Habermas, both truth and rightness are discursively vindicated but there is a significant difference. For truth, discourse merely recognizes or signifies that a statement is (believed to be) true in respect of an objective world. For morality, discourse actually justifies or creates the norm as a norm within the social world (Habermas 1999a, p. 38).
5. There are several versions of both U and D.
6. 'Fiduciary' generally means trustworthy, so a fiduciary remark is one that is said sincerely and is trustworthy.
7. My alignment of these categorizations is different from that usually assumed, for example in the *Dictionary of Critical Realism* (Hartwig 2007, Table 19) which correlates evidential with alethic and descriptive with epistemological.

11 Conclusion

In this book, we have been on a long journey covering many fields and disciplines. The aims set out in the Introduction were twofold: to provide a book that covered the main philosophical issues underlying the field of systems thinking that might in some ways be an update of Laszlo's book of 1972; and to demonstrate that both critical realism and the philosophy of science share many concepts that themselves draw on systems thinking, although this is largely unacknowledged within the disciplines. I hope that this book has gone some way towards achieving these aims. Certainly we have covered many of the major philosophical issues – ontology, epistemology, ethics, the philosophy of science and social science, emergence, causality, knowledge, truth, and methodology – that are of great relevance for systems thinking. And I hope that I have shown the confluence of ideas between critical realism and the philosophy of science around the idea of mechanisms and mechanistic explanation. My aspiration is that this will lead to a fertile engagement between all three disciplines.

Looking back to the 1960s when Laszlo's book was born, we find that science, especially social science, and systems thinking were dominated by positivism and functionalism and by analytic philosophy. During the 1970s, there was a major revolution that challenged positivism – variously called interpretivism, constructivism or the linguistic turn – which promised to move into new directions. This it did, especially in the form of soft systems or second-order cybernetics, but this ultimately resulted in a standoff between the two sides with somewhat of a chasm in between. This was only made worse by the advent of postmodernism, which threatened an end to philosophy itself. By then, any ontological contact with the external world was almost completely lost. It was into this battlefield that CR stepped and it has become so influential because it promises a rapprochement between positivism and constructivism, and a resurrection of realism.

I think we are now in a much better place to begin again to realize the promise of systems thinking. This is vital at the moment when the world has so many major problems – environmental, economic, cultural and religious. The one thing that perhaps most people can agree with is that these are all systemic problems. We are faced by challenges that are what Rittel (Rittel and Webber 1973) called 'wicked problems' – they involve the whole world, and are multi-layered and multifaceted. They can only be approached as a whole, in a systemic fashion, and this is why systems thinking should be the one approach that is fundamental to resolving these issues. But, as Chapter 10 suggested, it is also the case that this can only be achieved by people, all the people, recognizing that there are moral and environmental problems which ultimately have to transcend individual ethical, religious or nationalistic viewpoints.

Looking to the future, I believe that we are in a much stronger position now than we have been for many years, provided that we are:

Realist in acknowledging the actuality of an independent, causally efficacious world while recognizing the limitations on our access to it.

Systemic and interdisciplinary because the world is a complex intertwining or lamination of many kinds of mechanisms – physical, biological, psychological, social, political, economic – that interact in complex non-linear ways.

Empirical or evidence-based in accepting the importance of data and information and its analysis, especially when there is so much of it available – 'big data' and analytics – while recognizing the limitations of both the data itself and purely empiricist analyses of it.

Interpretive in accepting that in the social world individuals and groups to some extent construct their own interpretations and valuations, and that we must understand and pay due attention to this in researching and in resolving problems.

Multimethodological in that the above points imply that we need to be eclectic in our use of methods and methodologies in both research and in practical interventions, and much more interdisciplinary in our use of theory.

Critical and committed in recognizing the unavoidable ethical and moral dimensions to all our decisions and actions, and not hiding behind technocratic, managerialist or positivist arguments that they are somehow 'value-free'.

If we follow these maxims sincerely and consistently, then the systems approach and critical realism can genuinely help us tackle the intractable, wicked problems that currently beset the world.

Bibliography

Abbott, A. 1998. 'The Causal Devolution', *Sociological Methods and Research* (27:2), pp. 148–181.

Ackoff, R., and Emery, F. 1972. *On Purposeful Systems*. London: Tavistock.

Ackroyd, S., and Fleetwood, S. 2000. *Realist Perspectives on Management and Organisations*. London: Routledge.

Allen, G.P. and Fildes, R. 2000. 'Econometric Forecasting', in: *Principles of Forecasting: A Handbook for Researchers and Practitioners*, J.S. Armstrong (ed.). Norwell, MA: Kluwer, pp. 303–362.

Alvesson, M. and Karreman, D. 2001. 'Odd Couple: Making Sense of the Curious Concept of Knowledge Management', *Journal of Management Studies* (38:7), pp. 995–1018.

Alvesson, M. and Willmott, H. (eds). 1992. *Critical Management Studies*. London: SAGE Publications.

American Society for Cybernetics. 2006. 'A Timeline for the Evolution of Cybernetics', www.asc-cybernetics.org/foundations/timeline.htm, accessed 14 November 2006.

Anderson, P., Arrow, K. and Pines, D. (eds). 1988. *The Economy as an Evolving Complex System*. Boulder, CO: Westview Press.

Anderson, P., Blatt, R., Christianson, M., Grant, A., Marquis, C., Neuman, E., Sonenshein, S. and Sutcliffe, K. 2006. 'Understanding Mechanisms in Organizational Research', *Journal of Management Inquiry* (15:2), pp. 102–113.

Angel, I. 1997. 'Welcome to the 'Brave New World'', in: *Information Systems: An Emerging Discipline?*, J. Mingers and F. Stowell (eds). London: McGraw Hill, pp. 363–384.

Angus, J.E. and Clark, A.M. 2012. 'Using Critical Realism in Nursing and Health Research: Promise and Challenges', *Nursing Inquiry* (19:1), pp. 1–3.

Archer, M. 1988. *Culture and Agency*. Cambridge: Cambridge University Press.

Archer, M. 1990. 'Human Agency and Social Structure: A Critique of Giddens', in: *Anthony Giddens: Consensus and Controversy*, J. Clark, C. Modgil and S. Modgil (eds). Cambridge: Cambridge University Press, pp. 47–56.

Archer, M. 1995. *Realist Social Theory: The Morphogenetic Approach*. Cambridge: Cambridge University Press.

Archer, M. 1996. 'Social Integration and System Integration: Developing the Distinction', *Sociology* (30:4), pp. 679–699.

Archer, M. 1998. 'Realism and Morphogenesis', in: *Critical Realism: Essential Readings*, M. Archer, R. Bhaskar, A. Collier, T. Lawson and A. Norrie (eds). London: Routledge, pp. 356–382.

Archer, M. 2000a. *Being Human: The Problem of Agency*. Cambridge: Cambridge University Press.

Archer, M. 2000b. 'For Structure: Its Reality, Properties and Powers: A Reply to Anthony King', *Sociological Review* (48), pp. 464–472.

Archer, M., Bhaskar, R., Collier, A., Lawson, T. and Norrie, A. (eds). 1998. *Critical Realism: Essential Readings*. London: Routledge.

Aristotle. 2000. *Nicomachean Ethics*. Cambridge: Cambridge University Press.

Arthur, W.B. 1994. 'Positive Feedbacks in the Economy', *McKinsey Quarterly* (1), pp. 81–95.

Ashby, W.R. 1956. *An Introduction to Cybernetics*. London: Chapman Hall.

Astley, W. 1985. 'Administrative Science as Socially Constructed Truth', *Administrative Science Quarterly* (30), pp. 497–513.

Baert, P. 1996. 'Realist Philosophy of the Social Sciences and Economics: A Critique', *Cambridge Journal of Economics* (20), pp. 513–522.

Barnes, S. 1977. *Interests and the Growth of Knowledge*. London: Routledge.

Baron, M., Pettit, P. and Slote, M. 1997. *Three Methods of Ethics: A Debate*. Oxford: Blackwell.

Bartlett, S. and Suber, P. 1987. *Self Reference: Reflections on Reflexivity*. Dordrecht: Martinus Nijhoff.

Bateson, G. 1973a. 'Form, Substance and Difference', in: *Steps to an Ecology of Mind*, G. Bateson (ed.). London: Paladin, pp. 423–440.

Bateson, G. 1973b. *Steps to an Ecology of Mind*. Frogmore, Herts: Granada Publishing.

Bawden, R. 2007. 'Complexity – Unruly and Otherwise', *Perspectives in Biology and Medicine* (50:4), pp. 614–624.

Bechtel, W. 2001. 'The Compatibility of Complex Systems and Reduction: A Case Analysis of Memory Research', *Minds and Machines* (11:4), pp. 483–502.

Bechtel, W. and Abrahamsen, A. 2005. 'Explanation: A Mechanist Alternative', *Studies in History and Philosophy of Biological and Biomedical Sciences* (36:2), pp. 421–441.

Beck, U. 2000. *What Is Globalization?* Cambridge: Polity Press.

Bedau, M. 1997. 'Weak Emergence', *Philosophical Perspectives* (11), pp. 375–399.

Beer, S. 1966. *Decision and Control*. London: Wiley.

Beer, S. 1975. 'Preface.', in: *Autopoiesis and Cognition: The Realization of the Living*, H.M. Maturana and F.G. Varela (eds). Dordrecht: Reidel.

Beer, S. 1985. *Diagnosing the System for Organizations*. Chichester: Wiley.

Beinhocker, E. 1997. 'Strategy at the Edge of Chaos', *McKinsey Quarterly* (1), pp. 24–29.

Bell, D. 1999. 'The Axial Age of Technology: Foreword 1999', in: *The Coming of the Post-Industrial Society*, D. Bell (ed.). New York: Basic Books, pp. ix–lxxxv.

Bennett, P., Ackerman, F., Eden, C. and Williams, T. 1997. 'Analysing Litigation and Negotiation: Using a Combined Methodology', in: *Multimethodology: The Theory and Practice of Combining Management Science Methodologies*, J. Mingers and A. Gill (eds). Chichester: Wiley, pp. 59–88.

Benson, S. and Standing, C. 2001. 'Effective Knowledge Management: Knowledge, Thinking and the Personal-Corporate Knowledge Nexus Problem', *Information Systems Frontiers* (3:2), pp. 227–238.

Bentham, J. 1948 (orig. 1789). *Introduction to the Principles of Morals and Legislation*. Oxford: Oxford University Press.

Berger, P. and Luckman, T. 1967. *The Social Construction of Reality*. London: Penguin.

Berkeley, G. 1995 (orig. 1710). *A Treatise Concerning the Principles of Human Knowledge*. Indianapolis: Hackett Publishing Company.

Berry, D. 1996. *Statistics: A Bayesian Perspective*. Belmont, CA: Duxbury Press.

Beschorner, T. 2006. 'Ethical Theory and Business Practices: The Case of Discourse Ethics', *Journal of Business Ethics* (66), pp. 127–139.

Best, S. and Kellner, D. 1991. *Postmodern Theory: Critical Interrogations*. New York: Guilford Press.

Bhaskar, R. 1978. *A Realist Theory of Science*. Hemel Hempstead: Harvester.

Bhaskar, R. 1979. *The Possibility of Naturalism*. Hemel Hempstead: Harvester.

Bhaskar, R. 1986. *Scientific Realism and Human Emancipation*. London: Verso.

Bhaskar, R. 1989. *Reclaiming Reality*. London: Verso.

Bhaskar, R. 1991. *Philosophy and the Idea of Freedom* Oxford: Blackwell.

Bhaskar, R. 1993. *Dialectic: The Pulse of Freedom*. London: Verso.

Bhaskar, R. 1994. *Plato Etc*. London: Verso.

Bhaskar, R. 1997. 'On the ontological Status of Ideas', *Journal for the Theory of Social Behaviour* (27:2/3), pp. 139–147.

Bhaskar, R. 2000. *From East to West: Odyssey of a Soul*. London: Routledge.

Bhaskar, R. 2002a. 'Critical Realism and Ethnomethodology: Debate with Rom Harré', in: *From Science to Emancipation*, R. Bhaskar (ed.). London: Sage, pp. 96–124.

Bhaskar, R. (ed.). 2002b. *From Science to Emancipation: Alienation and the Actuality of Enlightenment*. London: Sage.

Bhaskar, R. 2002c. *The Philosophy of Meta-Reality: Creativity, Love and Freedom*. London: Routledge.

Bhaskar, R. 2002d. *Reflections on Meta-Reality: Transcendence, Emancipation and Everyday Life*. London: Routledge.

Bhaskar, R. 2010. 'Contexts of Interdisciplinarity: Interdisciplinarity and Climate Change', in: *Interdisciplinarity and Climate Change: Transforming Knowledge and Practice for Our Global Future*, R. Bhaskar, C. Frank, K.G. Høyer, P. Naess and J. Parker (eds). London: Routledge, pp. 1–24.

Bhaskar, R. 2013. 'Prolegomenon', in: *Engaging with the World: Agency, Institutions, Historical Formations*, M. Archer and A. Maccarini (eds). London: Routledge, pp. 11–21.

Bhaskar, R. and Hartwig, M. 2010. *The Formation of Critical Realism: A Personal Perspective*. London: Routledge.

Bijker, W., Hughes, T. and Pinch, T. (eds). 1987. *The Social Construction of Technological Systems*. Cambridge, MA: MIT Press.

Bird, A. 2008. 'Causal Exclusion and Evolved Emergent Properties', in: *Revitalizing Causality: Realism About Causality in Philosophy and Social Science*, R. Groff (ed.). London: Routledge, pp. 163–178.

Blackler, F. 1995. 'Knowledge, Knowledge Work and Organizations: An Overview and Interpretation', *Organization Studies* (16:6), pp. 16–36.

Bleicher, J. 1980. *Contemporary Hermeneutics*. London: Routledge.

Bloor, D. 1976. *Knowledge and Social Imagery*. London: Routledge.

Bogdanov, A. 1980 (orig. 1922). *Essays in Tektology: The General Science of Organization*. Seaside, CA: Intersystems Publications.

Boisot, M. 1995. *Information Space: A Framework for Learning in Organizations*. London: Routledge.

Boulding, K. 1956. 'General Systems Theory – the Skeleton of Science', *Management Science* (2:3), pp. 197–208.

Bowen, H. 1953. *Social Responsibilities of the Businessman*. New York: Harper.

Bowie, N. 1999. *Business Ethics: A Kantian Perspective*. Oxford: Blackwell.

Bradley, F. 1914. *Essays on Truth and Reality*. Oxford: Oxford University Press.

Brans, M. and Rossbach, S. 1997. 'The Autopoiesis of Administrative Systems: Niklas Luhmann on Public Administration and Public Policy', *Public Administration* (75:3), pp. 417–439.

Broad, D. 1925. *The Mind and Its Place in Nature*. London: Routledge & Keegan Paul.

Brocklesby, J. and Cummings, S. 1996. 'Foucault Plays Habermas: An Alternative Philosophical Underpinning for Critical Systems Thinking', *Journal of the Operational Research Society* (47:6), pp. 741–754.

Brown, A., Fleetwood, S. and Roberts, J.M. (eds). 2002. *Critical Realism and Marxism*. London: Routledge.

Brown, C. 1988. 'A New Interdisciplinary Impulse and the Anthropology of the 1990s', *International Social Science Journal* (40:2), pp. 211–220.

Bryant, G. 1995. *Practical Sociology*. Cambridge: Polity Press.

Bryman, A. 2001. *Social Research Methods*. Oxford: Oxford University Press.

Bryman, A. (ed.). 2006. *Mixed Methods*. London: Sage.

Bryson, J., Ackermann, F., Eden, C. and Finn, D. 2004. *Visible Thinking: Unlocking Causal Mapping for Practical Business Results*. Chichester: Wiley.

Buckley, W. 1967. *Sociology and Modern Systems Theory*. Englewood Cliffs, NJ: Prentice Hall.

Buckley, W. (ed.). 1968. *Modern Systems Research for the Behavioural Scientist*. Chicago: Aldine.

Bunge, M. 1992. 'System Boundary', *International Journal of General Systems* (20), pp. 215–229.

Bunge, M. 2004a. 'Clarifying some Misunderstandings about Social Systems and their Mechanisms', *Philosophy of the Social Sciences* (34:3), pp. 371–381.

Bunge, M. 2004b. 'How Does It Work? The Search for Explanatory Mechanisms', *Philosophy of the Social Sciences* (34:2), pp. 182–210.

Burrell, G. and Morgan, G. 1979. *Sociological Paradigms and Organisational Analysis*. London: Heinemann.

Byrne, D. 1998. *Complexity Theory and the Social Sciences*. London: Routledge.

Callinicos, A. 1995. 'Critical Realism and Beyond: Roy Bhaskar's Dialectic', University of York Department of Politics, Working Paper 7.

Cannon, W. 1939. *The Wisdom of the Body*. New York: Norton.

Capra, F. 1997. *The Web of Life: A New Synthesis of Mind and Matter*. London: Flamingo.

Cartwright, N. 1983. *How the Laws of Physics Lie*. Oxford: Clarenden Press.

Cartwright, N. 1997. 'Where do the Laws of Nature Come From?', *Dialectica* (51:1), pp. 65–78.

Cartwright, N. 1999. *The Dappled World*. Cambridge: Cambridge University Press.

Casati, R. and Varzi, A. 2002. 'Events', in: *Stanford Encyclopedia of Philosophy*. Stanford, CA: Stanford University.

Casella, G. and Berger, R. 1990. *Statistical Inference*. Belmont, CA: Wadsworth.

Center for Disease Control. 1996. 'Outbreak of Primary and Secondary Syphilis – Baltimore City, Maryland, 1995', *Morbidity and Mortality Weekly Report* (45:8), pp. 166–169.

Chakravartty, A. 2005. 'Causal Realism: Events and Processes', *Erkenntnis* (63:1), pp. 7–31.

Chakravartty, A. 2007. *A Metaphysics for Scientific Realism: Knowing the Unobservable*. Cambridge: Cambridge University Press.

Chakravartty, A. 2008. 'Inessential Aristotle: Powers without Essences', in: *Revitalizing Causality: Realism About Causality in Philosophy and Social Science*, R. Groff (ed.). London: Routledge, pp. 152–162.
Chalmers, A. 1988. 'Is Bhaskar's Realism Realistic?', *Radical Philosophy* (49:Summer), pp. 18–23.
Checkland, P. 1972. 'Towards a Systems-Based Methodology for Real-World Problem Solving', *Journal of Systems Engineering* (3:2), pp. 87–116.
Checkland, P. 1981. *Systems Thinking, Systems Practice*. Chichester: Wiley.
Checkland, P. 1983. 'O.R. and the Systems Movement – Mappings and Conflicts', *Journal of the Operational Research Society* (34:8), pp. 661–675.
Checkland, P. 1985. 'From Optimizing to Learning: A Development of Systems Thinking for the 1990s', *Journal of the Operational Research Society* (36:9), pp. 757–768.
Checkland, P. 1989. 'O.R. and Social Science: Fundamental Thoughts', *Operational Research and the Social Sciences*, M. Jackson, P. Keys and S. Cropper (eds), Cambridge: Plenum Press, pp. 35–41.
Checkland, P. 1999a. 'Soft Systems Methodology: A 30-Year Retrospective', in: *Systems Thinking, Systems Practice* (2nd edn), P. Checkland (ed.). Chichester: Wiley, pp. A1–A66.
Checkland, P. 1999b. *Systems Thinking, Systems Practice: Includes a 30-Year Retrospective* (2nd edn). Chichester: Wiley.
Checkland, P. and Holwell, S. 1998. *Information, Systems and Information Systems: Making Sense of the Field*. Chichester: Wiley.
Checkland, P. and Poulter, J. 2006. *Learning for Action*. Chichester: Wiley.
Checkland, P. and Scholes, J. 1990. *Soft Systems Methodology in Action*. Chichester: Wiley.
Chomsky, N. 1957. *Syntactic Structures*. The Hague: Mouton.
Chow, S. 1996. *Statistical Significance: Rationale, Validity and Utility*. New York: Sage.
Chow, S. 1998. 'Precis of *Statistical Significance: Rationale, Validity, and Utility*', *Behavioural and Brain Sciences* (21), pp. 169–239.
Cicourel, A. 1973. *Cognitive Sociology: Language and Meaning in Social Interaction*. London: Penguin.
Cilliers, P. 2000. *Complexity and Postmodernism: Understanding Complex Systems*. London: Routledge.
Clark, A. and Chalmers, D. 1998. 'The Extended Mind', *Analysis* (58:1), pp. 7–19.
Clarke, S. 2010. 'Transcendental Realisms in the Philosophy of Science: On Bhaskar and Cartwright', *Synthese* (173), pp. 299–315.
Cohen, I. 1989. *Structuration Theory: Anthony Giddens and the Constitution of Social Life*. London: Macmillan.
Cohen, J. 1994. 'The Earth Is Round (P<0.05)', *American Psychologist* (49:12), pp. 997–1003.
Collins, H. 1985. *Changing Order: Replication and Induction in Scientific Practice*. Beverly Hills, CA: Sage.
Collins, H. 1993. 'The Structure of Knowledge', *Social Research* (60), pp. 95–116.
Compte, A. 1986. *Auguste Comte: The Positive Philosophy of Auguste Comte*. London: George Bell & Son.
Cooley, T. and LeRoy, S. 1985. 'Atheoretical Macroeconomics: A Critique', *Journal of Monetary Economics* (16:3), pp. 283–308.
Cortina, J. and Dunlap, W. 1997. 'On the Logic and Purpose of Significance Testing', *Psychological Methods* (2:2), pp. 161–172.

Cox, K.R. 2013. 'Notes on a Brief Encounter: Critical Realism, Historical Materialism and Human Geography', *Dialogues in Human Geography* (3:1), pp. 3–21.

Craib, I. 1992. *Anthony Giddens*. London: Routledge.

Craver, C. 2006. 'When Mechanistic Models Explain', *Synthese* (153:3), pp. 355–376.

Craver, C. 2007. *Explaining the Brain*. Oxford: Clarendon Press.

Creswell, J. 2009. *Research Design: Quantitative, Qualitative and Mixed Method Approaches* (3rd ed.). Los Angeles: Sage.

Cruickshank, J. 2007. 'The Usefulness of Fallibilism in Post-Positivist Philosophy', *Philosophy of the Social Sciences* (37:3), pp. 263–288.

Cruickshank, J. 2012. 'Positioning Positivism, Critical Realism and Social Constructionism in the Health Sciences: A Philosophical Orientation', *Nursing Inquiry* (19:1), pp. 71–82.

Danermark, B., Ekstrom, M., Jakobsen, L. and Karlsson, J. 2002. *Explaining Society: Critical Realism in the Social Sciences*. London: Routledge.

Darden, L. 2006. *Reasoning in Biological Discoveries*. Cambridge: Cambridge University Press.

Davenport, T. and Prusak, L. 1998. *Working Knowledge*. Cambridge, MA: Harvard University Press.

David, P. 1986. 'Understanding the Economics of Qwerty', in: *Economic History and the Modern Economist*, W. Parker (ed.). Oxford: Blackwell, pp. 30–49.

Dawid, A. 2000. 'Causal Inference without Counterfactuals', *Journal of the American Statistical Association* (95), pp. 407–448.

Dawid, A.P. 2002. 'Influence Diagrams for Causal Modelling and Inference', *International Statistical Review/Revue Internationale de Statistique* (70:2), pp. 161–189.

de Graaf, G. 2006. 'Discourse and Descriptive Business Ethics', *Business Ethics: A European Review* (15:3), pp. 246–258.

Deetz, S. 1996. 'Describing Differences in Approach to Organization Science: Rethinking Burrell and Morgan and Their Legacy', *Organization Science* (7:2), pp. 191–207.

DeForge, R. and Shaw, J. 2012. 'Back- and Fore-Grounding ontology: Exploring the Linkages between Critical Realism, Pragmatism, and Methodologies in Health and Rehabilitation Sciences', *Nursing Inquiry* (19:1), pp. 83–95.

Derrida, J. 1978. *Writing and Difference*. London: RKP.

Dewey, J. 1938. *Logic: The Theory of Inquiry*. New York: Holt.

Donaldson, T. and Dunfee, T.W. 1995. 'Integrative Social Contracts Theory: A Communitarian Conception of Economic Ethics', *Economics and Philosophy* (11:1), April, pp. 85–112.

Donaldson, T. and Preston, L. 1995. 'The Stakeholder Theory of the Corporation: Concepts, Evidence and Implications', *Academy of Management Review* (20), pp. 65–91.

Dretske, F. 1981. *Knowledge and the Flow of Information*. Oxford: Blackwell.

Dreyfus, H. 1992. *What Computers Still Can't Do: A Critique of Artificial Reason* (3rd ed.). Cambridge, MA: MIT Press.

Driesch, H. 1908. *The Science and Philosophy of the Organism*. London: Black.

Dryzek, J. 2002. *Deliberative Democracy and Beyond: Liberals, Critics, Contestations*. Oxford: Oxford University Press.

Durkheim, E. 1951 (orig. 1897). *Suicide: A Study in Sociology*. Glencoe, IL: The Free Press.

Earl, M. 1994. 'Knowledge as Strategy: Reflections on Skandia International and Shorko Films', in: *Strategic Information Systems: A European Perspective*, C. Ciborra and T. Jelassi (eds). Chichester: Wiley.

Easterby Smith, M., Thorpe, R. and Jackson, P. 2008. *Management Research* (3rd ed.). London: Sage.

Eden, C. 1989. 'Using Cognitive Mapping for Strategic Options Development and Analysis', in: *Rational Analysis for a Problematic World*, J. Rosenhead (ed.). Chichester: Wiley, pp. 21–42.

Eden, C. 1993. 'From the Playpen to the Bombsite: The Changing Nature of Management Science', *Omega-International Journal of Management Science* (21:2), pp. 139–154.

Eden, C. and Ackerman, F. 1998. *Making Strategy: The Journey of Strategic Management*. London: Sage.

Elder-Vass, D. 2005. 'Emergence and the Realist Account of Cause', *Journal for Critical Realism* (4:2), pp. 315–338.

Elder-Vass, D. 2007a. 'For Emergence: Refining Archer's Account of Social Structure', *Journal for the Theory of Social Behaviour* (37:1), pp. 25–44.

Elder-Vass, D. 2007b. 'Luhmann and Emergentism: Competing Paradigms for Social Systems Theory?', *Philosophy of the Social Sciences* (37:4), pp. 408–432.

Elder-Vass, D. 2010. *The Causal Power of Social Structures*. Cambridge: Cambridge University Press.

Evan, W. and Freeman, R. 1988. 'A Stakeholder Theory of the Modern Corporation: Kantian Capitalism', in: *Ethical Theory and Business*, T. Beauchamp and N. Bowie (eds). Englewood Cliffs, NJ: Prentice Hall, pp. 75–93.

Faucheux, C. and Makridakis, S. 1979. 'Automation or Autonomy in Organizational Design', *International Journal of General Systems* (5), pp. 213–220.

Featherstone, M., Turner, B. and Hepworth, M. (eds). 1991. *The Body: Social Process and Cultural Theory*. London: Sage.

Fiaz, N. 2013. 'Constructivism Meets Critical Realism: Explaining Pakistan's State Practice in the Aftermath of 9/11', *European Journal of International Relations*, 4 January 2013.

Fildes, R. 1985. 'Quantitative Forecasting – the State of the Art: Econometric Models', *Journal of the Operational Research Society* (36:7), pp. 549–580.

Fildes, R. 2000. 'Book Review', *International Journal of Forecasting* (16:1), pp. 132–133.

Fildes, R. and Stekler, H. 2000. 'The State of Macroeconomic Forecasting', George Washington University Center for Economic Research Discussion Paper No. 99-04.

Fine, B. 2002. 'Addressing the Critical and the Real in Critical Realism', http://users.ox.ac.uk/~dialogue/documents/ben_fine_critreal.htm, accessed 14 April 2003.

Fleck, J. 1997. 'Contingent Knowledge and Technology Development', *Technology Analysis and Strategic Management* (9:4), pp. 383–397.

Fleetwood, S. (ed.). 1999. *Critical Realism in Economics: Development and Debate*. London: Routledge.

Fleetwood, S. 2001. 'Causal Laws, Functional Relations and Tendencies', *Review of Political Economy* (13:2), pp. 201–220.

Fleetwood, S. 2002. 'Boylan and O'Gorman's Causal Holism: A Critical Realist Evaluation', *Cambridge Journal of Economics* (26:1), pp. 27–45.

Fleetwood, S. 2009. 'The ontology of Things, Properties and Powers', *Journal of Critical Realism* (8:3), pp. 343–366.

Flegg, G. 1974. *From Geometry to Topology*. London: English Universities Press.

Flood, R. and Jackson, M. (eds). 1991. *Critical Systems Thinking: Directed Readings*. Chichester: Wiley.

Forrester, J. 1961. *Industrial Dynamics*. Cambridge, MA: MIT Press.

Forrester, J. 1969. *Urban Dynamics*. Cambridge, MA: MIT Press.

Forrester, J. 1973. *World Dynamics*. Cambridge, MA: MIT Press.

Foucault, M. 1980. *Power/Knowledge: Selected Interviews and Other Writings 1972–1977*. Brighton: Harvester Press.

Freeman, L. 2001. 'Information Systems Knowledge: Foundations, Definitions and Applications', *Information Systems Frontiers* (3:2), pp. 249–266.

Freeman, R. 1999. 'Divergent Stakeholder Theory', *Academy of Management Review* (24:2), pp. 233–236.

Freeman, R. and Phillips, R. 2002. 'Stakeholder Theory: A Libertarian Defence', *Business Ethics Quarterly* (12:3), pp. 331–349.

Frege, G. 1952. *Translations from the Philosophical Writings of Gottlob Frege*. Oxford: Blackwell.

French, S. and Smith, J. (eds). 1997. *The Practice of Bayesian Analysis*. London: Arnold.

Friedman, M. 1962. *Capitalism and Freedom*. Chicago: University of Chicago Press.

Friend, J. 2001. 'The Strategic Choice Approach', in: *Rational Analysis for a Problematic World Revisited: Problem Structuring Methods for Complexity, Uncertainty and Conflict*, J. Rosenhead and J. Mingers (eds). Chichester: Wiley, pp. 115–150.

Fuenmayor, R. 1991. 'Truth and Openness: An Epistemology for Interpretative Systemology', *Systems Practice* (4), pp. 473–490.

Fukuyama, F. 1992. *The End of History and the Last Man*. New York: Free Press.

Garcia-Marzá, D. 2005. 'Trust and Dialogue: Theoretical Approaches to Ethics Auditing', *Journal of Business Ethics* (57), pp. 209–219.

Gardner, M. and Altman, D. 1986. 'Confidence Intervals Rather Than P-Values. Estimation Rather Than Hypothesis Testing', *British Medical Journal* (292), pp. 746–750.

Garriga, E. and Melé, D. 2004. 'Corporate Social Responsibility: Mapping the Territory', *Journal of Business Ethics* (53:1–2), pp. 51–71.

Gell-Mann, M. 1995. *The Quark and the Jaguar: Adventures in the Simple and the Complex*. London: Abacus.

Gergen, K. 1999. *An Invitation to Social Construction*. London: Sage.

Gerring, J. 2007. 'Review Article: The Mechanismic Worldview: Thinking Inside the Box', *British Journal of Political Science* (38), pp. 161–179.

Gettier, E. 1963. 'Is Justified True Belief Knowledge?', *Synthesis* (23), pp. 121–123.

Gibbons, M., Limoges, L., Nowotny, H., Schwartman, S., Scott, P. and Trow, M. 1994. *The New Production of Knowledge: The Dynamics of Science and Research in Contemporary Societies*. London: Sage.

Giddens, A. (ed.). 1974. *Positivism and Sociology*. London: Heineman.

Giddens, A. 1976. *New Rules of Sociological Method*. London: Hutchinson.

Giddens, A. 1979. *Central Problems in Social Theory: Action, Structure and Contradiction in Social Analysis*. London: Macmillan.

Giddens, A. 1981. *A Contemporary Critique of Historical Materialism*. London: Macmillan Press.

Giddens, A. 1984. *The Constitution of Society*. Cambridge: Polity Press.

Giddens, A. 1987. *Social Theory and Modern Sociology*. Cambridge: Polity Press.

Giddens, A. 1989. 'A Reply to My Critics', in: *Social Theory of Modern Societies: Anthony Giddens and His Critics*, D. Held and J. Thompson (eds). Cambridge: Cambridge University Press, pp. 249–301.

Giddens, A. 1990a. *The Consequences of Modernity*. Cambridge: Polity Press.

Giddens, A. 1990b. 'Structuration Theory and Sociological Analysis', in: *Anthony Giddens: Consensus and Controversy*, J. Clark, C. Modgil and S. Modgil (eds). London: Falmer Press, pp. 297–315.

Giddens, A. 1992. *The Transformation of Intimacy: Sexuality, Love and Eroticism in Modern Society*. Cambridge: Polity Press.

Gillett, C. 2007. 'The Metaphysics of Mechanisms and the Challenge of the New Reductionism', in: *The Matter of the Mind,* M. Schouton and H. de Jong (eds). Oxford: Blackwell, pp. 74–100.

Gioia, D. 2003. 'Give It Up! Reflections on the Interpreted World (a Commentary on Meckler and Baillie)', *Journal of Management Inquiry* (12:3), pp. 285–292.

Gladwell, M. 2001. *The Tipping Point: How Little Things Can Make a Big Difference*. London: Abacus.

Gladwell, M. 2008. *Outliers: The Story of Success*. New York: Little, Brown and Co.

Gleick, J. 1988. *Chaos: Making a New Science*. London: Heinemann.

Glennan, S. 1996. 'Mechanisms and the Nature of Causation', *Erkenntnis* (44), pp. 49–71.

Glennan, S. 2002. 'Rethinking Mechanistic Explanation', *Philosophy of Science* (69:S3), pp S342–S353.

Goles, T. and Hirschheim, R. 2000. 'The Paradigm Is Dead, the Paradigm Is Dead.... Long Live the Paradigm: The Legacy of Burrell and Morgan', *Omega* (28:3), pp. 249–268.

Gorski, P.S. 2013. ''What Is Critical Realism? And Why Should You Care?'', *Contemporary Sociology: A Journal of Reviews* (42:5), pp. 658–670.

Greene, J. 2007. *Mixed Methods in Social Enquiry*. San Francisco: Jossey Bass.

Gregory, R. 1972. *Eye and Brain*. London: Weidenfeld.

Grey, C., Knights, D. and Willmott, H. 1996. 'Is a Critical Pedagogy of Management Possible?', in: *Rethinking Management Education*, R. French and C. Grey (eds). London: SAGE Publications.

Groff, R. 2000. 'The Truth of the Matter – Roy Bhaskar's Critical Realism and the Concept of Alethic Truth', *Philosophy of the Social Sciences* (30:3), pp. 407–435.

Groff, R. (ed.). 2008. *Revitalizing Causality: Realism About Causality in Philosophy and Social Science*. London: Routledge.

Groff, R. 2011. 'Getting Past Hume in the Philosophy of Social Science', in: *Causality in the Sciences*, P. Illari and J. Williamson (eds). Oxford: Oxford University Press.

Grosz, E. 1994. *Volatile Bodies: Towards a Corporeal Feminism*. Bloomington: Indiana University Press.

Grover, V. and Davenport, T. 2001. 'General Perspectives on Knowledge Management: Fostering a Research Agenda', *Journal of Management Information Systems* (18:1), pp. 5–21.

Haavelmo, T. 1943. 'The Statistical Implications of a System of Simultaneous Equations', *Econometrics* (11), pp. 1–12.

Habermas, J. 1971. *Towards a Rational Society*. London: Heinemann.

Habermas, J. 1974. *Theory and Practice*. London: Heinemann.

Habermas, J. 1976. *Legitimation Crisis*. London: Heinemann.

Habermas, J. 1978. *Knowledge and Human Interests* (2nd ed.). London: Heinemann.

Habermas, J. 1979. *Communication and the Evolution of Society*. London: Heinemann.

Habermas, J. 1984. *The Theory of Communicative Action*, vol. 1: *Reason and the Rationalization of Society*. London: Heinemann.

Habermas, J. 1985. 'Excursus on Luhmann's Appropriation of the Philosophy of the Subject through Systems Theory', in: *The Philosophical Discourse of Modernity*, J. Habermas (ed.). Cambridge: Polity Press, pp. 368–385.

Habermas, J. 1987. *The Theory of Communicative Action*, vol. 2: *Lifeworld and System: A Critique of Functionalist Reason*. Oxford: Polity Press.

Habermas, J. 1990. *Moral Consciousness and Communicative Action*. Cambridge: Polity Press.

Habermas, J. 1992a. 'Discourse Ethics, Law and Sittlichkeit', in: *Autonomy and Solidarity: Interviews with Jürgen Habermas*, P. Dews (ed.). London: Verso, pp. 245–271.

Habermas, J. 1992b. 'Discourse Ethics: Notes on a Programme of Philosophical Justification', in: *Moral Consciousness and Communicative Action*, J. Habermas (ed.). Cambridge: Polity Press, pp. 43–115.

Habermas, J. 1993a. *Justification and Application*. Cambridge: Polity Press.

Habermas, J. 1993b. 'On the Pragmatic, the Ethical, and the Moral Employments of Practical Reason', in: *Justification and Application*, J. Habermas (ed.). Cambridge: Polity Press, pp. 1–17.

Habermas, J. 1994. 'What Theories Can Accomplish – and What They Can't', in: *The Past as Future: Jurgen Habermas Interviewed by Michael Haller*, M. Haller (ed.). Cambridge: Polity Press, pp. 99–120.

Habermas, J. 1996. *Between Facts and Norms*. Cambridge: Polity Press.

Habermas, J. 1999a. 'A Genealogical Analysis of the Cognitive Content of Morality', in: *The Inclusion of the Other*. Cambridge: Polity Press, pp. 3–48.

Habermas, J. 1999b. *The Inclusion of the Other*. Cambridge: Polity Press.

Habermas, J. 2001. *The Postnational Constellation: Political Essays*. Cambridge: Polity Press.

Habermas, J. 2003a. 'Rightness Versus Truth', in: *Truth and Justification*, J. Habermas (ed.). Cambridge: Polity Press, pp. 238–275.

Habermas, J. 2003b. *Truth and Justification*. Cambridge: Polity Press.

Habermas, J. and Luhmann, N. 1971. *Theorie Der Gesellschaft Oder Sozialtechnologie*. Frankfurt: Suhrkamp.

Haeckel, E. 1866. *Generelle Morphologie Der Organismen*. Berlin: Verlag.

Hagen, R. 1997. 'In Praise of the Null Hypothesis Statistical Test', *American Psychologist* (52:1), pp. 15–24.

Hall, A.D. 1962. *A Methodology for Systems Engineering*. New York: Van Nostrand Reinhold.

Hall, N. 2004. 'Two Concepts of Causation', in: *Causation and Counterfactuals*, L. Paul, E. Hall and J. Collins (eds). Cambridge, MA: MIT Press, pp. 225–276.

Hammersley, M. 2002. 'Research as Emancipatory: The Case of Bhaskar's Critical Realism', *Journal of Critical Realism* (1:1), pp. 33–48.

Hansen, N. 1958. *Patterns of Discovery*. New York: Cambridge University Press.

Harman, G. 1965. 'The Inference to the Best Explanation', *The Philosophical Review* (74:1), pp. 88–95.

Harré, R. 1986. *Varieties of Realism*. Oxford: Blackwell.

Harré, R. 2002. 'Social Reality and the Myth of Social Structure', *European Journal of Social Theory* (5:1), pp. 111–123.

Harré, R. and Madden, E. 1975. *Causal Powers: A Theory of Natural Necessity*. Oxford: Blackwell.

Hartwig, M. 2007. *Dictionary of Critical Realism*. Abingdon: Routledge.

Harvey, D.L. 2002. 'Agency and Community: A Critical Realist Paradigm', *Journal for the Theory of Social Behaviour* (32:2), pp. 163–194.

Hayles, N. 1992. 'The Materiality of Informatics', *Configurations* (1), pp. 147–170.

Hayles, N.K. 1999. 'The Second Wave of Cybernetics: From Reflexivity to Self-Organization', in: *How We Became Posthuman: Virtual Bodies in Cybernetics, Literature, and Informatics*, N.K. Hayles (ed.). Chicago: University of Chicago Press, pp. 131–159.

Heims, S. 1993. *Constructing a Social Science for Postwar America: The Cybernetics Group 1946–1953*. Cambridge, MA: MIT Press.

Heisenberg, W. 1963. *Physics and Philosophy*. London: Allen and Unwin.

Hempel, C. 1965. *Aspects of Scientific Explanation*. New York: Free Press.

Hendry, D. 1995. *Dynamic Econometrics*. Oxford: Oxford University Press.

Hendry, D., Leamer, E. and Poirier, D. 1990. 'The ET Dialogue: A Conversation on Econometric Methodology', *Econometric Theory* (6), pp. 171–261.

Herborth, B. 2012. 'Theorising Theorising: Critical Realism and the Quest for Certainty', *Review of International Studies* (38:01), pp. 235–251.

Hesse, M. 1974. *The Structure of Scientific Inference*. Berkeley: University of California Press.

Hoover, K. 1988. *The New Classical Macroeconomics: A Sceptical Enquiry*. Oxford: Blackwell.

Horwich, P. 1991. *Truth*. Oxford: Blackwell.

Hostettler, N. and Norrie, A. 2003. 'Are Critical Realist Ethics Foundationalist?', in: *Critical Realism: The Difference It Makes*, J. Cruickshank (ed.). Abingdon: Routledge, pp. 30–53.

Houston, S. 2010. 'Prising Open the Black Box: Critical Realism, Action Research and Social Work', *Qualitative Social Work* (9:1), pp. 73–91.

Hughes, P. and Brecht, G. 1978. *Vicious Circles and Infinity*. Harmondsworth: Penguin.

Hume, D. 1967 (orig. 1750). *Enquiries Concerning Human Understanding and the Principles of Morals*. Oxford: Clarendon Press.

Hunt, S. 2005. 'For Truth and Realism in Management Research', *Journal of Management Inquiry* (14:2), pp. 127–138.

Husserl, E. 1973 (orig. 1913). *Logical Investigations*. London: Routledge.

Husserl, E. 1977 (orig. 1931). *Cartesian Meditations*. The Hague: Martinus Nijhoff.

Hutchins, E. 1995. *Cognition in the Wild*. Cambridge, MA: MIT Press.

Hutchison, T. 1994. 'Ends and Means in the Methodology of Economics', in: *New Directions in Economic Methodology*, R. Backhouse (ed.). London: Routledge.

Illari, P. and Williamson, J. 2011. 'Mechanisms Are Real and Local', in: *Causality in the Sciences*, P. Illari and J. Williamson (eds). Oxford: Oxford University Press, pp. 818–844.

Jackson, M. 1982. 'The Nature of Soft Systems Thinking: The Work of Churchman, Ackoff, and Checkland', *Journal of Applied Systems Analysis* (9), pp. 17–27.

Jackson, M. 1985. 'Social Systems Theory and Practice: The Need for a Critical Approach', *International Journal of General Systems* (10), pp. 135–151.

Jackson, M. 1991. 'The Origins and Nature of Critical Systems Thinking', *Systems Practice* (4:2), pp. 131–149.

Jackson, M. 2000. *Systems Approaches to Management*. Dordrecht: Kluwer Academic.

Jackson, M. 2003. *Systems Thinking: Creative Holism for Managers*. Chichester: Wiley.

Jakobson, R. and Halle, M. 1956. *Fundamentals of Language*. The Hague: Mouton.

James, W. 1976. *The Meaning of Truth*. Cambridge, MA: Harvard University Press.

Kant, I. 1991 (orig. 1785). *The Moral Law: Groundwork of the Metaphysic of Morals*. London: Routledge.

Kaufmann, S. 1995. *At Home in the Universe: The Search for the Laws of Complexity*. London: Penguin.

Kay, J. 1995. 'Cracks in the Crystal Ball', in: *Financial Times*, 29 September.

Keat, R. and Urry, J. 1981. *Social Theory as Science* (2nd ed.). London: RKP.

Kelemen, M. and Rumens, N. 2008. *An Introduction to Critical Management Research*. London: Sage.

Kickert, W. 1993. 'Autopoiesis and the Science of Public Administration: Essence, Sense and Nonsense', *Organization Studies* (14:2), pp. 261–278.

Kim, J. 1998. *Mind in a Physical World: An Essay on the Mind-Body Problem and Mental Causation*. Boston: MIT Press.

Kim, J. 2006. 'Emergence: Core Ideas and Issues', *Synthese* (151), pp. 547–559.

King, A. 1999a. 'Against Structure: A Critique of Morphogenetic Social Theory', *Sociological Review* (47:2), pp. 199–227.

King, A. 1999b. 'The Impossibility of Naturalism: The Antinomies of Bhaskar's Realism', *Journal for the Theory of Social Behaviour* (29:3), pp. 267–290.

King, A. 2000. 'The Accidental Derogation of the Lay Actor: A Critique of Giddens's Concept of Structure', *Philosophy of the Social Sciences* (30:3), pp. 362–383.

Klein, H. and Huynh, M. 2004. 'The Critical Social Theory of Jürgen Habermas and its Implications for IS Research', in: *Social Theory and Philosophy for Information Systems*, J. Mingers and L. Willcocks (eds). Chichester: Wiley, pp. 157–237.

Klein, P. 1971. 'A Proposed Definition of Propositional Knowledge', *The Journal of Philosophy* (68:16), pp. 471–482.

Knorr-Cetina, K., and Mulkay, M. (eds). 1983. *Science Observed*. London: Sage.

Koopmans, T. 1947. 'Measurement without Theory', *Review of Economics and Statistics* (29:2), pp. 161–172.

Korzybski, I. 1933. *Science and Sanity: An Introduction to Non-Aristotelian Systems and General Semantics*. Fort Worth, TX: Institute of General Semantics.

Kuhn, T. 1970. *The Structure of Scientific Revolutions*. Chicago: University of Chicago Press.

Kuhn, T. 1977. *The Essential Tension: Selected Studies in Scientific Tradition and Change*. Chicago: University of Chicago Press.

Lacey, H. 1997. 'Neutrality in the Social Sciences: On Bhaskar's Argument for an Essential Emancipatory Impulse in Social Science', *Journal for the Theory of Social Behaviour* (27:2–3), pp. 213–243.

LaFollette, H. (ed.). 2007. *Ethics in Practice*. Oxford: Blackwell.

Lakatos, I. 1978. *The Methodology of Scientific Research Programmes: Philosophical Papers Volume 1*. Cambridge: Cambridge University Press.

Lakoff, G. and Johnson, M. 1987. *Women, Fire and Dangerous Things: What Categories Reveal About the Mind*. Chicago: University of Chicago Press.

Laszlo, E. 1972. *Introduction to Systems Philosophy: Toward a New Paradigm of Contemporary Thought*. New York: Gordon and Breach.

Latour, B. 1987. *Science in Action*. Milton Keynes: Open University Press.

Laudan, L. 1984. *Science and Values: The Aims of Science and Their Role in Scientific Debate*. Berkeley, CA: University of California Press.

Lawson, T. 1997. *Economics and Reality*. London: Routledge.

Lawson, T. 2013. 'What Is This 'School' Called Neoclassical Economics?', *Cambridge Journal of Economics* (37:5), pp. 947–983.

Layder, D. 1985. 'Power, Structure and Agency', *Journal for the Theory of Social Behaviour* (15:2), pp. 131–149.

Layder, D. 1987. 'Key Issues in Structuration Theory: Some Critical Remarks', *Current Perspectives in Social Theory* (8), pp. 25–46.

Layder, D. 1989. 'The Macro/Micro Distinction, Social Relations and Methodological Bracketing: Unresolved Issues in Structuration Theory', *Current Perspectives in Social Theory* (9), pp. 123–141.

Layder, D. 1993. *New Strategies in Social Research*. Cambridge: Polity Press.

Layder, D. 1994. *Understanding Social Theory*. London: Sage.

Leamer, E. 1983. 'Let's Take the Con out of Econometrics', *American Economic Review* (73:1), pp. 31–43.

Lee, H. 1961. *Symbolic Logic*. New York: Random House.

Lee, M.-D. 2008. 'A Review of the Theories of Corporate Social Responsibility', *International Journal of Management Reviews* (10:1), pp. 53–73.

Lettvin, J., Maturana, H., McMulloch, W. and Pitts, W. 1959. 'What the Frog's Eye Tells the Frog's Brain', *Proceedings of the Institute for Radio Engineers* (47:11), pp. 1940–1951.

Levi-Strauss, C. 1963. *Structural Anthropology*. New York: Basic Books.

Levy, D. 1994. 'Chaos Theory and Strategy: Theory, Application, and Managerial Implications', *Strategic Management Journal* (15), pp. 167–178.

Lewes, G. 1875. *The Problems of Life and Mind*. London: Keegan Paul.

Lewin, R. 1992. *Complexity: Life at the Edge of Chaos*. New York: Macmillan.

Lewin, R. and Regine, B. 1999. *The Soul at Work*. London: Orion.

Lewis, P. 2000. 'Realism, Causality and the Problem of Social Structure', *Journal for the Theory of Social Behaviour* (30:3), pp. 249–268.

Lipton, P. 2004. *Inference to the Best Explanation* (2nd ed.). London: Routledge.

Liu, T. 1960. 'Under-Identification, Structural Estimation, and Forecasting', *Econometrica* (28), pp. 855–865.

Locke, J. 1980 (orig. 1689). *Second Treatise of Government*. Indianapolis: Hackett Publishing Company.

Lounsbury, M. 2003. 'The Death of Organization Science', *Journal of Management Inquiry* (12:3), pp. 293–298.

Luhmann, N. 1982a. *The Differentiation of Society*. New York: Columbia University Press.

Luhmann, N. 1982b. 'The World Society as a Social System', *International Journal of General Systems* (8), pp. 131–138.

Luhmann, N. 1984. *Soziale Systeme*. Frankfurt: Suhrkamp.

Luhmann, N. 1985a. 'The Self-Reproduction of Law and Its Limits', in: *Dilemmas of Law in the Welfare State*, G. Teubner (ed.). Berlin: de Gruyter.

Luhmann, N. 1985b. 'Society, Meaning, Religion – Based on Self-Reference', *Sociological Analysis* (46:1), pp. 5–20.

Luhmann, N. 1986. 'The Autopoiesis of Social Systems', in: *Sociocybernetic Paradoxes*, F. Geyer and J. van der Zouwen (eds). London: SAGE Publications.

Luhmann, N. 1987a. 'Closure and Openness: On Reality in the World of Law', in: *Autopoiesis and the Law*, G. Teubner (ed.). Berlin: de Gruyter, pp. 335–348.

Luhmann, N. 1987b. 'The Unity of the Legal System', in: *Autopoiesis and the Law*, G. Teubner (ed.). Berlin: de Gruyter, pp. 12–35.

Luhmann, N. 1989a. *Ecological Communication*. Cambridge: Polity Press.

Luhmann, N. 1989b. 'Law as a Social System', *Northwestern University Law Review* (83), pp. 136–150.

Luhmann, N. 1990a. *Essays in Self-Reference*. New York: Columbia University Press.

Luhmann, N. 1990b. 'Meaning as Sociology's Basic Concept', in: *Essays in Self-Reference*, N. Luhmann (ed.). New York: Columbia University Press, pp. 21–79.

Luhmann, N. 1992. 'Operational Closure and Structural Coupling: The Differentiation of the Legal System', *Cardozo Law Review* (13:5), pp. 1419–1441.

Luhmann, N. 1995. *Social Systems*. Stanford: Stanford University Press.

Luhmann, N. 1996. 'Membership and Motives in Social Systems', *Systems Research* (13:3), pp. 341–348.

Luhmann, N. 2000a. *Organisation und Entscheidung*. Opladen: Verlag.

Luhmann, N. 2000b. *The Reality of the Mass Media*. Cambridge: Polity Press.

Machamer, P. 2004. 'Activities and Causation: The Metaphysics and Epistemology of Mechanisms', *International Studies in the Philosophy of Science* (18:1), pp. 27–39.

Machamer, P., Darden, L. and Craver, C.F. 2000. 'Thinking About Mechanisms', *Philosophy of Science* (67:1), pp. 1–25.

Machlup, F. 1980. *Knowledge and Knowledge Production*. Princeton: Princeton University Press.

MacIntyre, A. 1985. *After Virtue*. London: Duckworth.

Mackay, D. 1969. *Information, Mechanism and Meaning*. Cambridge, MA: MIT Press.

Mackie, J. 1977. 'Dispositions, Grounds and Causes', *Synthese* (34), pp. 361–370.

Mainzer, K. 1997. *Thinking in Complexity: The Complex Dynamics of Matter, Mind, and Mankind* (3rd ed.). Berlin: Springer.

Mandelbrot, B. 1982. *The Fractal Geometry of Nature*. New York: W.H. Freeman.

Manicas, P. 1998. 'A Realist Social Science', in: *Critical Realism: Essential Readings*, M. Archer, R. Bhaskar, A. Collier, T. Lawson and A. Norrie (eds). London: Routledge, pp. 313–338.

Manicas, P. 2006. *A Realist Philosophy of Social Science*. Cambridge: Cambridge University Press.

Mann, P. 2004. *Introductory Statistics* (5th ed.). New York: Wiley.

Marquis, J.-P. 1996. 'A Critical Note on Bunge's 'System Boundary' and a New Proposal', *International Journal of General Systems* (24:3), pp. 245–255.

Marsden, R. 1993. 'The Politics of Organizational Analysis', *Organization Studies* (14:1), pp. 93–124.

Marshall, N. and Sapsed, J. 2000. 'The Limits of Disembodied Knowledge: Challenges of Inter-Project Learning in the Production of Complex Products and Systems', paper presented at the conference 'Knowledge Management: Concepts and Controversies', University of Warwick, Coventry, 10–11 February.

Maruyama, M. 1963. 'The Second Cybernetics: Deviation-Amplifying Mutual Causal Processes', *American Scientist* (51), pp. 164–179.

Matten, D. and Crane, A. 2005. 'Corporate Citizenship: Toward an Extended Theoretical Conceptualization', *Academy of Management Journal* (30), pp. 166–179.

Maturana, H. 1970a. 'Biology of Cognition', Biological Computer Laboratory Research Report 9.0, University of Illinois, Urbana.

Maturana, H. 1970b. 'The Neurophysiology of Cognition', in: *Cognition: A Multiple View*, P. Garvin (ed.). New York: Spartan Books, pp. 3–23.

Maturana, H. 1975a. 'Communication and Representation Functions', in: *Encyclopedie de la Pleiade, Series Methodique*, J. Piaget (ed.). Paris: Gallimard.

Maturana, H. 1975b. 'The Organization of the Living, a Theory of the Living Organization', *International Journal of Man-Machine Studies* (7), pp. 313–332.

Maturana, H. 1978a. 'Biology of Language: The Epistemology of Reality', in: *Psychology and Biology of Language and Thought: Essays in Honour of Eric Lenneberg*, G.A. Miller and E. Lenneberg (eds). New York: Academic Press, pp. 27–63.

Maturana, H. 1978b. 'Cognition', in: *Wahrnehmung Und Kommunikation*, P. Hejl, W. Kock and G. Roth (ed.). Frankfurt: Peter Lang, pp. 29–49.

Maturana, H. 1980a. 'Biology of Cognition', in: *Autopoiesis and Cognition: The Realization of the Living*, H. Maturana and F. Varela (eds). Dordrecht: Reidel, pp. 1–58.

Maturana, H. 1980b. 'Man and Society', in: *Autopoietic Systems in the Social Sciences*, F. Benseler, P. Hejl and W. Kock (eds). Frankfurt: Campus Verlag, pp. 11–31.

Maturana, H. 1988. 'Reality: The Search for Objectivity or the Quest for a Compelling Argument', *Irish Journal of Psychology* (9), pp. 25–82.

Maturana, H. 1990. 'Science and Daily Life: The ontology of Scientific Explanations', in: *Selforganization: Portrait of a Scientific Revolution*, W. Krohn, G. Kuers and H. Nowotny (eds). Dordrecht: Kluwer Academic Publishers, pp. 12–35.

Maturana, H. 1993. 'Biology of the Aesthetic Experience', in: *Zeichen (Theorie) in Der Praxis*, M. Titzmann (ed.). Passau: Wissenschaftsverlag Rothe, pp. 37–56.

Maturana, H. 1995. 'Biology of Self-Consciousness', in: *Consciousness: Distinction and Reflection*, G. Trautteur (ed.). Naples: Bibliopolis.

Maturana, H. and Varela, F. 1975. 'Autopoietic Systems – a Characterization of the Living Organization', Biological Computer Laboratory Research Report 9.4, University of Illinois, Urbana.

Maturana, H. and Varela, F. 1980. *Autopoiesis and Cognition: The Realization of the Living*. Dordrecht: Reidel.

Maturana, H. and Varela, F. 1987. *The Tree of Knowledge*. Boston: Shambhala.

Maturana, H., Lettvin, J., McCulloch, S. and Pitts, W. 1960. 'Anatomy and Physiology of Vision in the Frog', *Journal of General Physiology* (43), pp. 129–175.

Maturana, H., Mpodozis, J., and Letelier, J. 1995. 'Brain, Language and the Origin of Human Mental Functions', *Biological Research* (28), pp. 15–26.

Maturana, H., Uribe, G. and Frenk, S. 1968. 'A Biological Theory of Relativistic Colour Coding in the Primate Retina', *Archiva de Biologia y Medicina Experimentales* (Suplemento 1), pp. 1–30.

Mayntz, R. 2004. 'Mechanisms in the Analysis of Social Macro-Phenomena', *Philosophy of the Social Sciences* (34:2), pp. 237–259.

McClosky, D. and Ziliak, S. 1996. 'The Standard Error of Regressions', *Journal of Economic Literature* (34:1), pp. 97–114.

McCullough, W. and Pitts, W. 1943. 'A Logical Calculus of the Ideas Immanent in Nervous Activity', *Bulletin of Mathematical Biophysics* (5), pp. 115–133.

McGann, J. 1991. *The Textual Condition*. Princeton: Princeton University Press.

McLaughlin, B. 1997. 'Emergence and Supervenience', *Intellectica* (2), pp. 25–43.

McNaught, K. 2001. 'An Introduction to Bayesian Belief Networks', *OR 43*, J. Ranyard (ed.), Bath: Operational Research Society, pp. 39–62.

McWherter, D. 2012. 'Reconstructing Bhaskar's Transcendental Analysis of Experiemntal Activity', *Cosmos and History: The Journal of Natural and Social Philosophy* (8:1), pp. 199–227.

Meadows, D., Meadows, D., Randers, J. and Behrens, W. 1972. *The Limits to Growth*. London: Pan.

Meckler, M. and Baillie, J. 2003a. 'The Truth About Social Construction in Administrative Science', *Journal of Management Inquiry* (12:3), pp. 273–284.

Meckler, M. and Baillie, J. 2003b. 'You Can't Handle the Truth: A Response to Gioia and Lounsbury', *Journal of Management Inquiry* (12:3), pp. 299–303.

Meehl, P. 1997. 'The Problem Is Epistemology Not Statistics. Replace Significance Tests by Confidence Intervals and Quantify Accuracy of Risky Numerical Predictions', in: *What If There Were No Significance Tests?*, L. Harlow, S. Mulaik and J. Steiger (eds). Hillsdale: Erlbaum, pp. 391–423.

Meisenbach, R. 2006. 'Habermas's Discourse Ethics and Principle of Universalization as a Moral Framework for Organization Communication', *Management Communication Quarterly* (20:1), pp. 39–62.

Menary, R. (ed.). 2006. *Radical Enactivism: Intentionality, Phenomenology and Narrative*. Amsterdam: John Benjamins.

Mendoza, J. 1997. 'The Duality of Structure', in: *Anthony Giddens: Critical Assessments*, G. Bryant and D. Jary (eds). London: Routledge, pp. 219–270.

Merleau-Ponty, M. 1962. *Phenomenology of Perception*. London: Routledge.

Merleau-Ponty, M. 1969. *The Visible and the Invisible*. Evanston, IL: Northwestern University Press.

Meyering, T.C. 2000. 'Physicalism and Downward Causation in Psychology and the Special Sciences', *Inquiry: An Interdisciplinary Journal of Philosophy* (43:2), pp. 181–202.

Meynen, H.T. 1992. 'The Bringing Forth of Dialogue: Latour Versus Maturana', in: *New Perspectives on Cybernetics: Self-Organization, Autonomy and Connectionism*, G. van de Vijver (ed.). Dordrecht: Kluwer Academic, pp. 157–174.

Midgley, G. 1992. 'Pluralism and the Legitimation of Systems Science', *Systems Practice* (5: 2), pp. 147–172.

Midgley, G. 1995. 'What Is This Thing Called Critical Systems Thinking?', in: *Critical Issues in Systems Theory and Practice*, K. Ellis, A. Gregory, B. Mears-Young and G. Ragsdell (eds). New York: Plenum Press, pp. 61–71.

Midgley, G. 2000. *Systemic Intervention: Philosophy, Methodology, and Practice*. New York: Kluwer Plenum.

Mill, J. 1843. *System of Logic*. London: Longmans.

Millar, J., Dermaid, A. and Quintas, P. 1997. 'Trans-Organisational Innovation: A Framework for Research', *Technology Analysis and Strategic Management* (9:4), pp. 399–418.

Miller, K.D. and Tsang, E.W.K. 2011. 'Testing Management Theories: Critical Realist Philosophy and Research Methods', *Strategic Management Journal* (32:2), pp. 139–158.

Mingers, J. 1980. 'Towards an Appropriate Social Theory for Applied Systems Thinking: Critical Theory and Soft Systems Methodology', *Journal of Applied Systems Analysis* (7:April), pp. 41–50.

Mingers, J. 1984. 'Subjectivism and Soft Systems Methodology – a Critique', *Journal of Applied Systems Analysis* (11), pp. 85–103.

Mingers, J. 1989a. 'An Introduction to Autopoiesis: A Reply to Fenton Robb', *Systems Practice* (2:3), pp. 349–351.

Mingers, J. 1989b. 'Problems of Measurement', in: *Operational Research and the Social Sciences*, M. Jackson, P. Keys and S. Cropper (eds). New York: Plenum Press, pp. 471–477.

Mingers, J. 1992a. 'The Problems of Social Autopoiesis', *International Journal of General Systems* (21:2), pp. 229–236.

Mingers, J. 1992b. 'Recent Developments in Critical Management Science', *Journal of the Operational Research Society* (43:1), pp. 1–10.

Mingers, J. 1995a. 'Information and Meaning: Foundations for an Intersubjective Account', *Information Systems Journal* (5), pp. 285–306.

Mingers, J. 1995b. *Self-Producing Systems: Implications and Applications of Autopoiesis*. New York: Plenum Press.

Mingers, J. 1996a. 'A Comparison of Maturana's Autopoietic Social Theory and Giddens' Theory of Structuration', *Systems Research* (13:4), pp. 469–482.

Mingers, J. 1996b. 'An Evaluation of Theories of Information with Regard to the Semantic and Pragmatic Aspects of Information Systems', *Systems Practice* (9:3), pp. 187–209.

Mingers, J. 1997a. 'A Critical Evaluation of Maturana's Constructivist Family Therapy', *Systems Practice* (10:2), pp. 137–151.

Mingers, J. 1997b. 'Systems Typologies in the Light of Autopoiesis: A Reconceptualization of Boulding's Hierarchy, and a Typology of Self-Referential Systems', *Systems Research and Behavioural Science* (14), pp. 303–313.

Mingers, J. 1997c. 'Towards Critical Pluralism', in: *Multimethodology: Theory and Practice of Combining Management Science Methodologies*, J. Mingers and A. Gill (eds). Chichester: Wiley, pp. 407–440.

Mingers, J. 2000a. 'The Contribution of Critical Realism as an Underpinning Philosophy for OR/MS and Systems', *Journal of the Operational Research Society* (51:11), pp. 1256–1270.

Mingers, J. 2000b. 'An Idea Ahead of Its Time: The History and Development of Soft Systems Methodology', *Systemic Practice and Action Research* (13:6), pp. 733–756.

Mingers, J. 2000c. 'What Is It to Be Critical? Teaching a Critical Approach to Management Undergraduates', *Management Learning* (31:2), pp. 219–237.

Mingers, J. 2001a. 'Combining IS Research Methods: Towards a Pluralist Methodology', *Information Systems Research* (12:3), pp. 240–259.

Mingers, J. 2001b. 'Embodying Information Systems: The Contribution of Phenomenology', *Information and Organization* (11:2), pp. 103–128.

Mingers, J. 2002. 'Can Social Systems Be Autopoietic? Assessing Luhmann's Social Theory', *Sociological Review* (50:2), pp. 278–299.

Mingers, J. 2003. 'A Classification of the Philosophical Assumptions of Management Science Methods', *Journal of the Operational Research Society* (54:6), pp. 559–570.

Mingers, J. 2004a. 'Can Social Systems Be Autopoietic? Bhaskar's and Giddens' Social Theories', *Journal for the Theory of Social Behaviour* (34:4), pp. 403–426.

Mingers, J. 2004b. 'Paradigm Wars: Ceasefire Announced, Who Will Set up the New Administration?', *Journal Information Technology* (19), pp. 165–171.

Mingers, J. 2004c. 'Real-Izing Information Systems: Critical Realism as an Underpinning Philosophy for Information Systems', *Information and Organization* (14:2), pp. 87–103.

Mingers, J. 2006. 'A Critique of Statistical Modelling in Management Science from a Critical Realist Perspective: Its Role within Multimethodology', *Journal of the Operational Research Society* (55:2), pp. 202–219.

Mingers, J. 2008. 'Management Knowledge and Knowledge Management: Realism and Forms of Truth', *Knowledge Management Research and Practice* (6), pp. 62–76.

Mingers, J. 2009. 'Discourse Ethics and Critical Realist Ethics: An Evaluation in the Context of Business', *Journal of Critical Realism* (8:2), pp. 172–200.

Mingers, J. 2011. 'Ethics and OR: Operationalizing Discourse Ethics', *European Journal of Operational Research* (210:1), pp. 114–124.

Mingers, J. 2013. 'Can the Laws of Form Represent Syllogisms?', Kent Business School, Canterbury.

Mingers, J. and Brocklesby, J. 1997. 'Multimethodology: Towards a Framework for Mixing Methodologies', *Omega* (25:5), pp. 489–509.

Mingers, J. and Gill, A. (eds). 1997. *Multimethodology: Theory and Practice of Combining Management Science Methodologies*. Chichester: Wiley.

Mingers, J. and Walsham, G. 2008. 'Towards Ethical Information Systems: The Contribution of Discourse Ethics', in: *International Conference on Information Systems (ICIS2008)*. Paris: AIS.

Mingers, J. and White, L. 2010. 'A Review of the Recent Contribution of Systems Thinking to Operational Research and Management Science', *European Journal of Operational Research* (207:3), pp. 1147–1161.

Mitchell, G. 1993. *The Practice of Operational Research*. Chichester: Wiley.

Modell, S. 2009. 'In Defence of Triangulation: A Critical Realist Approach to Mixed Methods Research in Management Accounting', *Management Accounting Research* (20:3), pp. 208–221.

Morgan, G. 1986. *Images of Organisation*. Newbury Park: Sage.

Morrow, R. and Brown, D. 1994. *Critical Theory and Methodology*. London: Sage Publications.

Mosekilde, E. and Laugesen, L. 2007. 'Nonlinear Dynamic Phenomena in the Beer Model', *System Dynamics Review* (23:2/3), pp. 229–252.

Mumford, S. 2008. 'Powers, Dispositions, Properties or a Causal Realist Manifesto', in: *Revitalizing Causality: Realism About Causality in Philosophy and Social Science*, R. Groff (ed.). London: Routledge, pp. 139–151.

Murray, P. 1998. 'Complexity Theory and the Fifth Discipline', *Systems Practice* (11:3), pp. 275–292.

Myers, M.D. and Klein, H.K. 2011. 'A Set of Principles for Conducting Critical Research in Information Systems' *MIS Quarterly* (23:1), pp. 17–36.

Neurath, O. and McGuinness, B. 1987. *Unified Science: The Vienna Circle Monograph Series Originally Edited by Otto Neurath*. Dordrecht: D. Reidel.

New, C. 1994. 'Structure, Agency and Social Transformation', *Journal for the Theory of Social Behaviour* (24:3), pp. 187–205.

New, C. 1995. 'Sociology and the Case for Realism', *Sociological Review* (43:4), pp. 808–827.

Nickerson, R. 2000. 'Null Hypothesis Significance Testing: A Review of an Old and Continuing Controversy', *Psychological Methods* (5:2), pp. 241–301.

Nonaka, I. and Takeuchi, H. 1995. *The Knowledge-Creating Company: How Japanese Companies Create the Dynamics of Innovation*. Oxford: Oxford University Press.

O'Connor, T. and Wong, H. 2005. 'The Metaphysics of Emergence', *Noûs* (39:4), pp. 658–678.

O'Flynn, I. 2006. *Deliberative Democracy and Divided Societies*. Edinburgh: Edinburgh University Press.

Oliver, R., and Smith, J. (eds). 1990. *Influence Diagrams, Belief Nets, and Decision Analysis*. Chichester: Wiley.

Olsen, W. 1999. 'Developing Open-Systems Interpretations of Path Analysis: Fragility Analysis Using Farm Data from India', *Critical Realism: Implications for Practice*, Örebro University, Sweden: Centre for Critical Realism.

Onwuegbuzie, A. 2012. 'Putting the Mixed Back into Quantitative and Qualitative Research in Educational Research and Beyond: Moving Towards the Radical Middle', *International Journal of Multiple Research Approaches* (6:3), pp. 192–219.

Outhwaite, W. 1987. *New Philosophies of Social Science: Realism, Hermeneutics and Critical Theory*. London: Macmillan.

Palazzo, G. and Scherer, A. 2006. 'Corporate Legitimacy as Deliberation: A Communicative Framework', *Journal of Business Ethics* (66), pp. 71–88.

Parkinson, J. 2006. 'Deliberating in the Real World: Problems of Legitimacy in Deliberative Democracy.' Oxford: Oxford University Press.

Parsons, T. 1951. *The Social System*. Glencoe, IL: The Free Press.

Pask, G. 1976. *Conversation Theory*. New York: Elsevier.

Paterson, J. and Teubner, G. 1998. 'Changing Maps: Empirical Legal Autopoiesis', *Social and Legal Studies* (7:4), pp. 451–486.

Pawson, R. and Tilley, N. 1997. *Realistic Evaluation*. London: Sage.

Pearl, J. 2000. *Causality: Models, Reasoning, and Inference*. Cambridge: Cambridge University Press.

Pearl, J. 2009. 'Causal Inference in Statistics: An Overview', *Statistics Surveys* (3), pp. 96–146.

Peirce, C. 1878. 'How to Make Our Ideas Clear', *Popular Science Monthly* (12:January), pp. 286–302.

Peirce, C. 1931 1958. *Collected Papers of Charles Sanders Peirce* (8 volumes). Cambridge, MA: Harvard University Press.

Phillips, R. 2003. 'Stakeholder Legitimacy', *Business Ethics Quarterly* (13:1), pp. 25–41.
Piaget, J. 1969. *The Mechanisms of Perception*. New York: Basic Books.
Pickel, A. 2004. 'Systems and Mechanisms', *Philosophy of the Social Sciences* (34:2), pp. 169–181.
Pickel, A. 2007. 'Rethinking Systems Theory: A Programmatic Introduction', *Philosophy of the Social Sciences* (37:4), pp. 391–407.
Pickering, A. 2002. 'Cybernetics and the Mangle: Ashby, Beer and Pask', *Social Studies of Science* (32:3), pp. 413–437.
Pidd, M. 1996. *Tools for Thinking: Modelling in Management Science*. Chichester: Wiley.
Pilgrim, D. 2013. 'Some Implications of Critical Realism for Mental Health Research', *Social Theory and Health* (12), pp. 1–21.
Plano Clark, V. and Cresswell, J. 2007. *The Mixed Methods Reader*. Thousand Oaks, CA: Sage.
Plowright, D. 2011. *Using Mixed Methods: Frameworks for an Integrated Methodology*. Thousand Oaks, CA: Sage.
Polanyi, M. 1958. *Personal Knowledge: Towards a Post-Critical Philosophy*. London: Routledge.
Pole, A., West, M., and Harrison, J. 1994. *Applied Bayesian Forecasting and Time Series Analysis*. London: Chapman Hall.
Popper, K. 1959. *The Logic of Scientific Discovery*. London: Hutchinson.
Popper, K. 1962. *The Open Society and Its Enemies*. London: RKP.
Popper, K. 1969. *Conjectures and Refutations*. London: Routledge and Kegan Paul.
Popper, K. 1972. *Objective Knowledge: An Evolutionary Approach*. London: Oxford University Press.
Porpora, D. 1989. 'Four Concepts of Social Structure', *Journal for the Theory of Social Behaviour* (19:2), pp. 195–211.
Porpora, D. 1998a. 'Do Realist Run Regressions?', paper presented at the Second Conference of the International Centre for Critical Realism, University of Essex.
Porpora, D. 1998b. 'Four Concepts of Social Structure', in: *Critical Realism: Essential Readings*, M. Archer, R. Bhaskar, A. Collier, T. Lawson and A. Norrie (eds). London: Routledge, pp. 339–355.
Porter, M. 1985. *Competitive Advantage: Creating and Sustaining Superior Advantage*. New York: Free Press.
Prahalad, C., and Hammond, A. 2002. 'Serving the World's Poor, Profitably', *Harvard Business Review* (80:9), pp. 48–58.
Pratschke, J. 2003. 'Realistic Models? Critical Realism and Statistical Models in the Social Sciences', *Philosophica* (71), pp. 13–38.
Pratt, A. 1995. 'Putting Critical Realism to Work: The Practical Implications for Geographical Research', *Progress in Human Geography* (19:1), pp. 61–74.
Pratten, S. 2009. 'Critical Realism and Causality: Tracing the Aristotelian Legacy', *Journal for the Theory of Social Behaviour* (39:2), pp. 189–218.
Prigogine, I. and Stengers, I. 1984. *Order out of Chaos*. New York: Bantam.
Psillos, S. 2009. 'An Explorer Upon Untrodden Ground: Peirce on Abduction', in: *Handbook of the History of Logic: Inductive Logic*, D. Gabbay, S. Hartmann and J. Woods (eds). Amsterdam: Elsevier, pp. 117–151.
Putnam, H. 1975. *Mathematics, Matter and Method*. Cambridge: Cambridge University Press.
Putnam, H. 1981. *Reason, Truth, and History*. Cambridge: Cambridge University Press.
Quine, W. 1992. *Pursuit of Truth*. Boston: Harvard University Press.

Ramsey, F. 1927. 'Facts and Propositions', *Proceedings of the Aristotelian Society* (7), pp. 153–170.

Rawls, J. 1971. *A Theory of Justice*. Boston: Harvard University Press.

Reed, D. 1999a. 'Stakeholder Management Theory: A Critical Theory Perspective', *Business Ethics Quarterly* (9:3), pp. 453–483.

Reed, D. 1999b. 'Three Realms of Corporate Responsibility: Distinguishing Legitimacy, Morality and Ethics', *Journal of Business Ethics* (21), pp. 23–35.

Reed, M. 1997. 'In Praise of Duality and Dualism: Rethinking Agency and Structure in Organizational Analysis', *Organization Studies* (18:1), pp. 21–42.

Reed, M. 2001. 'Organization, Trust and Control: A Realist Analysis', *Organization Studies* (22:2), pp. 201–228.

Reichardt, C. and Gollob, H. 1997. 'When Confidence Intervals Should Be Used Instead of Statistical Tests, and Vice Versa', in: *What If There Were No Significance Tests?*, L. Harlow, S. Mulaik and J. Steiger (eds). Hillsdale: Erlbaum, pp. 259–284.

Rittel, H. and Webber, M. 1973. 'Dilemmas in a General Theory of Planning', *Policy Sciences* (4), pp. 155–169.

Ritter, W. 1919. *The Unity of the Organism, or the Organismal Conception of Life*. Boston: Gorham Press.

Robb, F. 1989a. 'Cybernetics and Suprahuman Autopoietic Systems', *Systems Practice* (2:1), pp. 47–74.

Robb, F. 1989b. 'The Limits to Human Organisation: The Emergence of Autopoietic Systems', in: *Operational Research and the Social Sciences*, M. Jackson, P. Keys and S. Cropper (eds). New York: Plenum Press, pp. 247–251.

Roberts, D. 1973. *The Existential Graphs of C.S. Peirce*. The Hague: Mouton.

Robey, D. 1996. 'Diversity in Information Systems Research: Threat, Promise and Responsibility', *Information Systems Research* (7:4), pp. 400–408.

Ron, A. 1999. 'Regression Analysis and the Philosophy of Social Sciences – a Critical Realist View', *Critical Realism: Implications for Practice*, Örebro University, Sweden: Centre for Critical Realism.

Rorty, R. 1980. *Philosophy and the Mirror of Nature*. Oxford: Blackwell.

Rorty, R. 1982. *Consequences of Pragmatism*. Minneapolis: University of Minnesota Press.

Rorty, R. 1989. *Contingency, Irony and Solidarity*. Cambridge: Cambridge University Press.

Rosenau, P. 1992. *Post-Modernism and the Social Sciences*. Princeton: Princeton University Press.

Rosenberg, A. 1992. *Economics: Mathematical Politics or Science of Diminishing Returns*. Chicago: University of Chicago Press.

Rosenblueth, A., Weiner, N. and Bigelow, J. 1943. 'Behaviour, Purpose and Teleology', *Philosophy of Science* (10), pp. 18–24.

Rosenhead, J. and Mingers, J. (eds). 2001. *Rational Analysis for a Problematic World Revisited*. Chichester: Wiley.

Russell, B. 1912. *The Problems of Philosophy*. Oxford: Oxford University Press.

Ryan, P. 2005. 'Meckler and Baillie on Truth and Objectivity', *Journal of Management Inquiry* (14:2), pp. 120–126.

Ryle, G. 1963. *The Concept of Mind*. London: Peregrine Books.

Salmon, W. 1998a. *Causality and Explanation*. Oxford: Oxford University Press.

Salmon, W. 1998b. 'Comets, Pollen and Dreams: Some Reflections on Scientific Explanation', in: *Causality and Explanation*. Oxford: Oxford University Press, pp. 50–67.

Saussure, F. de. 1960 (orig. 1916). *Course in General Linguistics*. London: Peter Owen.

Sawyer, R.K. 2004. 'The Mechanisms of Emergence', *Philosophy of the Social Sciences* (34:2), 2004, pp. 260–282.

Sayer, A. 1992. *Method in Social Science: A Realist Approach* (2nd ed.). London: Routledge.

Sayer, A. 1997a. 'Critical Realism and the Limits to Critical Social Science', *Journal For the Theory of Social Behaviour* (27:4), pp. 473–488.

Sayer, A. 1997b. 'Essentialism, Social Constructionism, and Beyond', *Sociological Review* (45:3), pp. 453–487.

Sayer, A. 2000. *Realism and Social Science*. London: Sage.

Sayer, A. 2013. 'Looking Forward to New Realist Debates', *Dialogues in Human Geography* (3:1), pp. 22–25.

Scherer, A. and Palazzo, G. 2007. 'Towards a Political Conception of Corporate Responsibility: Business and Society Seen from a Habermasian Perspective', *Academy of Management Review* (32:4), pp. 1096–1120.

Schlick, M. 1979. *Philosophical Papers*, H.L. Mulder and B.F.B. van de Velde-Schlick (eds), P. Heath (transl.). Dordrecht: D. Reidel.

Schultze, U. and Leidner, D. 2002. 'Studying Knowledge Management in Information Systems Research: Discourses and Theoretical Assumptions', *MIS Quarterly* (26:3), pp. 213–242.

Schultze, U. and Stabell, C. 2004. 'Knowing What You Don't Know? Discourses and Contradictions in Knowledge Management Research', *Journal of Management Studies* (41:4), pp. 549–573.

Schutz, A. 1972. *The Phenomenology of the Social World*. London: Heinemann.

Searle, J. 1996. *The Construction of Social Reality*. London: Penguin Books.

Senge, P. 1990. *The Fifth Discipline: The Art and Practice of the Learning Organization*. London: Century Books.

Sethi, S. 2002. 'Standards for Corporate Conduct in the International Arena: Challenges and Opportunities for Multinational Corporations', *Business and Society Review* (107), pp. 20–40.

Sewell, W. 1992. 'A Theory of Structure: Duality, Agency, and Transformation', *American Journal of Sociology* (98:1), pp. 1–29.

Shannon, C. and Weaver, W. 1949. *The Mathematical Theory of Communication*. Urbana, IL: University of Illinois Press.

Sheffer, H. 1913. 'Five Independent Postulates for Boolean Algebra', *Transactions of the American Mathematics Society* (14), pp. 481–488.

Sherden, W. 1998. *The Fortune Sellers: The Big Business of Buying and Selling Predictions*. New York: Wiley.

Shilling, C. 1993. *The Body and Social Theory*. London: Sage.

Shoemaker, S. 1997. 'Causality and Properties', in: *Properties*, D. Mellor and A. Oliver (eds). New York: Oxford University Press, pp. 228–254.

Simons, P. 1987. *Parts: A Study in ontology*. Oxford: Clarendon Press.

Singer, P. 1993. *Practical Ethics*. Cambridge: Cambridge University Press.

Skinner, B. 1976. *About Behaviorism*. New York: Vintage.

Smith, A. 2002 (orig. 1759). *The Theory of Moral Sentiments*. Cambridge: Cambridge University Press.

Smith, D. 1995. 'Tables Turn on Britain's Leading Forecasters', *Sunday Times*, 24 December.

Smith, J. 2004. 'A Precis of a Communicative Theory of the Firm', *Business Ethics: A European Review* (13:4), pp. 317–331.

Solomon, R. 1992. 'Corporate Roles, Personal Virtues: An Aristotelian Approach to Business Ethics', *Business Ethics Quarterly* (2:3), pp. 317–340.

Spash, C.L. 2012. 'New Foundations for Ecological Economics', *Ecological Economics* (77), pp. 36–47.

Spencer-Brown, G. 1972. *Laws of Form*. New York: Julien Press.

Spirtes, P., Glymour, C. and Scheines, R. 2001. *Causation, Prediction, and Search*. Boston: MIT Press.

Stacey, R. 2004. *Strategic Management & Organizational Dynamics: The Challenge of Complexity* (4th ed.). London: Pitman Publishing.

Steinmetz, G. 1998. 'Critical Realism and Historical Sociology: A Review Article', *Comparative Studies in Society and History* (40:1), pp. 170–186.

Stenmark, D. 2001. 'Leveraging Tacit Organizational Knowledge', *Journal of Management Information Systems* (17:3), pp. 9–24.

Sterman, J. 2000. *Business Dynamics: Systems Thinking and Modelling for a Complex World*. New York: McGraw-Hill.

Stern, R. 2000. *Transcendental Arguments and Scepticism: Answering the Question of Justification*. Oxford: Oxford University Press.

Stichweh, R. 2000. 'Systems Theory as an Alternative to Action Theory? The Rise of 'Communication' as a Theoretical Option', *Acta Sociologica* (43:1), pp. 5–13.

Stokes, P. 1990. 'Socio-Cybernetics and the Project of Scientificisation of Sociology', in: *Self-Steering and Cognition in Complex Systems: Towards a New Cybernetics*, E. Heylighen, E. Rosseel and F. Demeyere (eds). London: Gordon and Breech.

Strawson, P. 1950. 'Truth', *Proceedings of the Aristotelian Society* (24), pp. 129–156.

Suber, P. 1990. *Paradox of Self-Amendment*. New York: P. Lang Publishers.

Swan, J. and Scarbrough, H. 2001. 'Knowledge Management: Concepts and Controversies', *Journal of Management Studies* (38:7), pp. 913–921.

Syed, J., Mingers, J. and Murray, P. 2009. 'Beyond Rigour and Relevance: A Critical Realist Approach to Business Education', *Management Learning* (41:1), pp. 71–85.

Symons, J. 2008. 'Computational Models of Emergent Properties', *Minds and Machines* (18:4), pp. 475–491.

Synnott, A. 1993. *The Body Social: Symbolism, Self and Society*. London: Routledge.

Tarski, A. 1944. 'The Semantic Conception of Truth', *Philosophy and Phenomenological Research* (4), pp. 341–375.

Tashakkori, A. and Teddlie, C. 1998. *Mixed Methodology: Combining Qualitative and Quantitative Approaches*. London: SAGE Publications.

Tashakkori, A. and Teddlie, C. 2003. *Handbook of Mixed Methods in Social and Behavioural Research*. Thousand Oaks, CA: Sage.

Taylor, C. 1989. *The Sources of the Self: The Making of the Modern Identity*. Cambridge, MA: Harvard University Press.

Tell, F. 2004. 'What Do Organizations Know? Dynamics of Justification Contexts in R&D Activities', *Organization* (11:4), pp. 443–471.

Teubner, G. (ed.). 1987. *Autopoiesis and the Law*. Berlin: de Gruyter.

Teubner, G. 1993. *Law as an Autopoietic System*. Oxford: Blackwell.

Teubner, G. and Febbrajo, A. (eds). 1992. *State, Law and Economy as Autopoietic Systems: Regulation and Autonomy in a New Perspective*. Milan: Giuffre.

Thompson, E. and Varela, F.J. 2001. 'Radical Embodiment: Neural Dynamics and Consciousness', *Trends in Cognitive Sciences* (5:10), pp. 418–425.

Thompson, J. 1989. 'The Theory of Structuration', in: *Social Theory of Modern Societies: Anthony Giddens and his Critics*, D. Held and J. Thompson (eds). Cambridge: Cambridge University Press, pp. 56–76.

Thompson, M. and Walsham, G. 2004. 'Placing Knowledge Management in Context', *Journal of Management Studies* (41:5), pp. 725–747.

Tilly, C. 2004. 'Social Boundary Mechanisms', *Philosophy of the Social Sciences* (34:2), pp. 211–236.

Torres, P. 2009. 'A Modified Conception of Mechanisms', *Erkenntnis* (71), pp. 233–251.

Tsang, E. and Kwan, K. 1999. 'Replication and Theory Development in Organizational Science: A Critical Realist Perspective', *Academy of Management Review* (24:4), pp. 759–780.

Tsoukas, H. and Vladimirou, E. 2001. 'What Is Organizational Knowledge?', *Journal of Management Studies* (38:7), pp. 973–992.

Tuomi, I. 1999. 'Data Is More Than Knowledge: Implications of the Reversed Knowledge Hierarchy for Knowledge Management and Knowledge Memory', *Journal of Management Information Systems* (16:3), pp. 103–117.

Turner, B. 1984. *The Body and Society*. Oxford: Blackwell.

Turner, G. 2008. 'A Comparison of 'The Limits to Growth' with Thirty Years of Reality', Commonwealth Scientific and Industrial Research Organisation (CSIRO). Canberra.

Ulrich, W. 1994. *Critical Heuristics of Social Planning: A New Approach to Practical Philosophy* (2nd ed.). Chichester: Wiley.

Ulrich, W. 2000. 'Reflective Practice in the Civil Society: The Contribution of Critically Systemic Thinking', *Reflective Practice* (1:2), pp. 247–268.

van der Spek, R. and Spijkervet, A. 1997. 'Knowledge Management: Dealing Intelligently with Knowledge', in: *Knowledge Management and Its Integrative Elements*, J. Liebowitz and L. Wilcox (eds). New York: CRC Press, pp. 31–59.

van Fraassen, B. 1989. *Laws and Symmetry*. Oxford: Clarenden.

Vanderstraeten, R. 2000. 'Autopoiesis and Socialization: On Luhmann's Reconceptualization of Communication and Socialization', *British Journal of Sociology* (51:3), pp. 581–598.

Varela, C. 1999. 'Determinism and the Recovery of Human Agency: The Embodying of Persons', *Journal for the Theory of Social Behaviour* (29:4), pp. 385–402.

Varela, C. 2002. 'The Impossibility of Which Naturalism? A Response and a Reply', *Journal for the Theory of Social Behaviour* (32:1), pp. 105–111.

Varela, C. and Harré, R. 1996. 'Conflicting Varieties of Realism: Causal Powers and the Problems of Social Structure', *Journal For the Theory of Social Behaviour* (26:3), p. 313–25.

Varela, F. 1977. 'On Being Autonomous:The Lessons of Natural History for Systems Theory', in: *Applied Systems Research*, G. Klir (ed.). New York: Plenum Press, pp. 77–85.

Varela, F. 1979. *Principles of Biological Autonomy*. New York: Elsevier-North Holland.

Varela, F. 1981a. 'Autonomy and Autopoiesis', in: *Self-Organising Systems: An Interdisciplinary Approach*, G. Roth and H. Schwegler (eds). Frankfurt: Campus Verlag, pp. 14–24.

Varela, F. 1981b. 'Describing the Logic of the Living: The Adequacy and Limitations of the Idea of Autopoiesis', in: *Autopoiesis: A Theory of the Living Organization*, M. Zeleny (ed.). New York: Elsevier-North Holland, pp. 36–48.

Varela, F., Coutinho, A., Dupire, B. and Vaz, N. 1988. 'Cognitive Networks: Immune, Neural, and Otherwise', in: *Theoretical Immunology: SFI Series on the Science of Complexity*, A. Perelson (ed.). New York: Addison Wesley, pp. 359–375.

Varela, F., Thompson, E. and Rosch, E. 1991. *The Embodied Mind*. Cambridge, MA: MIT Press.

Veld, R.J., Schaap, L., Termeer, C. and Twist, M. 1991. *Autopoiesis and Configuration Theory: New Approaches to Societal Steering*. Dordrecht: Kluwer.

Venkatesh, V., Brown, S. and Bala, H. 2013. 'Bridging the Qualitative-Quantitative Divide: Guidelines for Conducting Mixed Methods in Information Systems', *MIS Quarterly* (37:1), pp. 21–54.

von Bertalanffy, L. 1950. 'The Theory of Open Systems in Physics and Biology', *Science* (111), pp. 23–29.

von Bertalanffy, L. 1971. *General Systems Theory*. London: Penguin.

Von Foerster, H. 1975. *The Cybernetics of Cybernetics*. Urbana, IL: Biological Computer Laboratory, University of Illinois.

Von Foerster, H. 1984. *Observing Systems*. Intersystems Publications.

Von Glasersfeld, E. 1984. 'An Introduction to Radical Constructivism.' in: *The Invented Reality.*, P. Watzlawick (ed.). New York: Norton.

von Krogh, G. and Grand, S. 2000. 'Justification in Knowledge Creation: Dominant Logic in Management Discourses', in: *Knowledge Creation: A Source of Value*, G. von Krogh, I. Nonaka and T. Nishiguchi (eds). Basingstoke: Macmillan, pp. 13–35.

von Krogh, G. and Roos, J. 1994. 'An Essay on Corporate Epistemology', *Strategic Management Journal* (15), pp. 53–71.

von Neumann, J. 1958. *The Computer and the Brain*. New Haven: Yale University Press.

von Uexküll, J. 1909. *Umwelt Und Innenwelt Der Tiere*. Berlin: Springer.

Walby, S. 2007. 'Complexity Theory, Systems Theory, and Multiple Intersecting Social Inequalities', *Philosophy of the Social Sciences* (37:4), pp. 449–470.

Waldrop, M. 1992. *Complexity: The Emerging Science at the Edge of Order and Chaos*. London: Viking.

Weiner, N. 1948. *Cybernetics: OR Communication and Control in the Animal and the Machine*. Cambridge, MA: MIT Press.

Werhane, P. and Freeman, R. 1999. 'Business Ethics: The State of the Art', *International Journal of Management Reviews* (1:1), pp. 1–16.

West, M. and Harrison, J. 1989. *Bayesian Forecasting and Dynamic Models*. New York: Springer-Verlag.

White, L. 1994. 'The Death of the Expert', *Journal of the Operational Research Society* (45:7), pp. 733–748.

White, L. and Taket, A. 1996. 'The End of Theory?', *Omega, International Journal of Management Science* (24:1), pp. 47–56.

Whitehead, A.N. and Russell, B. 1925. *Principia Mathematica* (2nd ed.). Cambridge: The University Press.

Whittington, R. 1992. 'Putting Giddens into Action – Social Systems and Managerial Agency', *Journal Management Studies* (29:6), pp. 693–712.

Wiig, K. 1993. *Knowledge Management Foundations: Thinking About Thinking – How People and Organizations Create, Represent, and Use Knowledge*. Arlington, TX: Schema Press.

Wilden, A. 1977. *System and Structure*. London: Tavistock.

Williamson, J. 2005. *Bayesian Nets and Causality*. Oxford: Oxford University Press.

Wimsatt, W. 1994. 'The ontology of Complex Systems: Levels, Perspectives and Causal Thickets', *Canadian Journal of Philosophy* (20), pp. 201–274.

Wimsatt, W. 2007. *Re-Engineering Philosophy for Limited Beings: Piecewise Approximations to Reality*. Harvard University Press.

Winograd, T. and Flores, F. 1987. *Understanding Computers and Cognition*. New York: Addison Wesley.

Winter, S. 1987. 'Knowledge and Competence as Strategic Assets', in: *The Competitive Challenge*, D. Teece (ed.). Cambridge, MA: Ballinger, pp. 159–184.

Witt, C. 2008. 'Aristotelian Powers', in: *Revitalizing Causality: Realism About Causality in Philosophy and Social Science*, R. Groff (ed.). London: Routledge, pp. 129–138.

Wittgenstein, L. 1958. *Philosophical Investigations*. Oxford: Blackwell.

Wittgenstein, L. 1974. *Tractatus Logico-Philosophicus*. London: Routledge and Kegan Paul.

Wolfe, C. 1995. 'In Search of Post-Humanist Theory: The Second-Order Cybernetics of Maturana and Varela', *Cultural Critique* (Spring), pp. 33–70.

Woodger, J. 1929. *Biological Principles: A Critical Study*. London: Keegan, Paul and Co.

Woodward, J. 2002. 'What Is a Mechanism? A Counterfactual Account', *Philosophy of Science* (69:S3), pp. S366–S377.

Woodward, J. 2003. *Making Things Happen: A Theory of Causal Explanation*. Oxford: Oxford University Press.

Woolgar, S. (ed.). 1988. *Knowledge and Reflexivity: New Frontiers in the Sociology of Knowledge*. Beverly Hills, CA: Sage.

Wright, C. 1999. 'They Shoot Dead Horses Don't They? Locating Agency in the Agent-Structure Problematique', *European Journal of International Relations* (5:1), pp. 109–142.

Wright, C. 2004. 'Theorising the Mechanisms of Conceptual and Semiotic Space', *Philosophies of the Social Sciences* (34), pp. 283–299.

Wright, S. 1921. 'Correlation and Causation', *Journal of Agricultural Research* (20), pp. 557–595.

Wynn, D. and Williams, C.K. 2012. 'Principles for Conducting Critical Realist Case Study Research in Information Systems', *MIS Quarterly* (36:3), pp. 787–810.

Yeung, H. 1997. 'Critical Realism and Realist Research in Human Geography', *Progress in Human Geography* (21:1), pp. 51–74.

Zachariadis, M., Scott, S. and Barrett, M. 2013. 'Methodological Implications of Critical Realism for Mixed-Methods Research', *MIS Quarterly* (37:3), pp. 855–879.

Zadeh, L.A. 1965. 'Fuzzy Sets', *Information and Control* (8:3), pp. 338–353.

Zeleny, M. and Hufford, C. 1992. 'The Application of Autopoiesis in Systems Analysis: Are Autopoietic Systems Also Social Systems?', *International Journal of General Systems* (21:2), pp. 145–160.

Zellner, A. 1997. *Bayesian Analysis in Econometrics and Statistics*. Cheltenham: Edward Elgar.

Index

Page numbers in *italics* denote tables, those in **bold** denote figures.

Taylor & Francis
eBooks
FOR LIBRARIES
ORDER YOUR FREE 30 DAY INSTITUTIONAL TRIAL TODAY!
Over 23,000 eBook titles in the Humanities, Social Sciences, STM and Law from some of the world's leading imprints.
Choose from a range of subject packages or create your own!
Benefits for you
Free MARC records
COUNTER-compliant usage statistics
Flexible purchase and pricing options
Benefits for your user
Off-site, anytime access via Athens or referring URL
Print or copy pages or chapters
Full content search
Bookmark, highlight and annotate text
Access to thousands of pages of quality research at the click of a button
For more information, pricing enquiries or to order a free trial, contact your local online sales team.
UK and Rest of World: online.sales@tandf.co.uk
US, Canada and Latin America:
e-reference@taylorandfrancis.com
www.ebooksubscriptions.com
ALPSP Award for BEST eBOOK PUBLISHER 2009 Finalist
Taylor & Francis eBooks
Taylor & Francis Group
A flexible and dynamic resource for teaching, learning and research.